Holt McDougal Mathematics

Course 3
Chapter 2 Resource Book

Printed in the United States of America

ISBN 13: 978-0-55-400726-7
ISBN 10: 0-55-400726-6

2 3 4 5 170 12 11 10

Contents

Description of Contents

Family Involvement Pages

The Chapter Resource Book includes a set of family involvement pages for each section. The family involvement pages consist of the following items:

- **Family Letter**, a two-page letter describing the math that the student will study in each section. A list of vocabulary words from the lessons is included, and some worked-out examples are provided.
- **At-Home Practice**, a one-page worksheet with problems drawn from the content described in the Family Letter. Answers are provided on the page so that the student and his or her family can check this work at home.
- **Family Fun**, a one-page activity sheet that the student and his or her family can work on together.

Practice A, B, and C

There are three practice worksheets for every lesson. All of these reinforce the content of the lesson. Practice B is shown in the Teacher's Edition and is appropriate for the on-level student. It is also available as a workbook (Homework & Practice Workbook).

Practice A is easier than Practice B but still practices the content of the lesson. Practice C is more challenging than Practice B.

Review for Mastery

The Review for Mastery worksheet (one per lesson) provides an alternate way to teach or review the main concepts of the lesson. This worksheet is one or two pages long and is shown in the Teacher's Edition.

Challenge

The Challenge worksheet (one per lesson) enhances critical thinking skills and extends the lesson. This worksheet is shown in the Teacher's Edition.

Problem Solving

The Problem Solving worksheet (one per lesson) provides practice in problem solving and opportunities for real-world applications and for interdisciplinary connections. There are both multiple choice and short response problems. This worksheet is shown in the Teacher's Edition.

Reading Strategies

The Reading Strategies worksheet (one per lesson) provides tools to help the student master math vocabulary or symbols.

Puzzles, Twisters & Teasers

The Puzzles, Twisters & Teasers worksheet (one per lesson) provides fun practice while reinforcing the content of the lesson.

CHAPTER 2

Family Letter

2A Rational Number Operations

Dear Family,

Up to this point, the student has been working with whole numbers and integers to solve equations and evaluate expressions. The student will be introduced to the set of **rational numbers.** This set of numbers includes fractions and decimals, which are used frequently outside the classroom.

A rational number is any number that can be written as a fraction $\frac{n}{d}$, where n and d are integers and $d \neq 0$. Decimals that repeat or terminate are also rational numbers.

The student will learn to simplify fractions, evaluate expressions using the four basic operations, and solve simple equations and inequalities using rational numbers.

Writing an answer in simplest form is a skill that the student must master. The goal of simplifying a fraction is to make sure the numerator and the denominator have no common factors other than 1. This is called being **relatively prime.**

Dividing the numerator and the denominator by the same common factor is one simple way of simplifying a fraction.

Simplify.

$\frac{30}{45}$

$30 = 2 \cdot 15$
$45 = 3 \cdot 15$ 15 is a common factor.

$\frac{30}{45} = \frac{30 \div 15}{45 \div 15}$

$\frac{2}{3}$ Divide the numerator and the denominator by 15.

The student will be asked to write answers in simplest form throughout the chapters to follow.

The student will also learn to compare and order rational numbers. A number line is useful in this process. The numbers on a number line run from least to greatest from left to right. To compare rational numbers, find each one on the number line. The number farthest to the right is greatest.

$\frac{1}{2}$ □ $\frac{1}{3}$

0 $\frac{1}{8}$ $\frac{1}{4}$ $\frac{1}{3}$ $\frac{1}{2}$ $\frac{2}{3}$ $\frac{3}{4}$ 1

Vocabulary

These are the math words we are learning:

least common denominator the least common multiple of two or more denominators

rational number any number that can be written as a fraction

reciprocals two numbers whose product is equal to one

relatively prime numbers that have no common factors other than 1

Family Letter

2A Rational Number Operations continued

$\frac{1}{2}$ is farther right than $\frac{1}{3}$ so $\frac{1}{2} > \frac{1}{3}$. You can use this technique to put a group of rational numbers in order from greatest to least or least to greatest.

The student will also learn to add, subtract, multiply, and divide rational numbers. When adding or subtracting fractions, there are some basic steps that must be followed.

- If the denominators are the same then add or subtract only the numerators and keep the denominator the same. Write the answer in simplest form.

$$\frac{7}{12} + \frac{3}{12} = \frac{7+3}{12} = \frac{10}{12} = \frac{5}{6}$$

- If the denominators are not the same, you must find a common denominator and then rewrite the fractions using the common denominator. Add or subtract as instructed.

$$\frac{3}{4} - \frac{1}{6} = \frac{3}{4}\left(\frac{3}{3}\right) - \frac{1}{6}\left(\frac{2}{2}\right) = \frac{9}{12} - \frac{2}{12} = \frac{7}{12}$$

To multiply fractions, multiply the numerators to get the numerator of the product and multiply the denominators to get the denominator of the product.

To divide fractions, multiply the dividend by the **reciprocal** of the divisor. To find the reciprocal, just flip the numerator and the denominator.

$$\frac{7}{8} \div \frac{2}{5} = \frac{7}{8} \bullet \frac{5}{2} = \frac{7 \bullet 5}{8 \bullet 2} = \frac{35}{16} = 2\frac{3}{16}$$

Discuss with the student the importance of rational numbers in everyday situations. Create problems that will allow the student to practice the skills learned.

Sincerely,

Name ______________________ Date ______________ Class ______________

CHAPTER 2

At-Home Practice

2A Rational Number Operations

Simplify.

1. $-\frac{18}{36}$ 2. $\frac{15}{35}$ 3. $-\frac{24}{36}$ 4. $\frac{25}{40}$

Compare. Write <, > or =.

5. $\frac{5}{2}$ ___ $2\frac{1}{2}$ 6. $\frac{7}{8}$ ___ $\frac{7}{16}$ 7. $\frac{34}{2}$ ___ 19

Order from least to greatest.

8. 0.25, $\frac{1}{8}$, $\frac{25}{5}$, 2.5 9. 1.3, $\frac{1}{3}$, $\frac{-4}{3}$, 3 10. −7, −5, −0.5, −0.725

Add or subtract.

11. $\frac{8}{12}-\frac{7}{12}$ 12. $\frac{7}{15}+\frac{4}{15}$ 13. $\frac{2}{7}+\frac{13}{21}$ 14. $\frac{-8}{9}+\frac{2}{3}$

15. $\frac{3}{4}+\frac{5}{8}$ 16. $\frac{5}{6}+\frac{3}{4}$ 17. $\frac{11}{15}-\frac{4}{5}$ 18. $1\frac{1}{4}-3\frac{1}{2}$

Multiply. Write each answer in simplest form.

19. $5\left(\frac{1}{7}\right)$ 20. $-6\left(1\frac{2}{3}\right)$ 21. $\frac{4}{5}\left(\frac{-3}{10}\right)$ 22. $\frac{7}{8}\left(\frac{-3}{5}\right)$

Divide. Write each answer in simplest form.

23. $\frac{1}{8}\div\frac{2}{3}$ 24. $\frac{3}{4}\div\frac{5}{4}$ 25. $\frac{-3}{4}\div\frac{2}{3}$ 26. $7\frac{1}{2}\div\left(-1\frac{1}{4}\right)$

Answers: 1. $-\frac{1}{2}$ **2.** $\frac{3}{7}$ **3.** $-\frac{2}{3}$ **4.** $\frac{5}{8}$ **5.** = **6.** > **7.** < **8.** $\frac{1}{8}$, 0.25, 2.5, $\frac{25}{5}$ **9.** $\frac{-4}{3}$, $\frac{1}{3}$, 1.3, 3 **10.** −7, −5, −0.725, −0.5 **11.** $\frac{1}{12}$ **12.** $\frac{11}{15}$ **13.** $\frac{19}{21}$ **14.** $-\frac{2}{9}$ **15.** $1\frac{3}{8}$ **16.** $1\frac{7}{12}$ **17.** $-\frac{1}{15}$ **18.** $-2\frac{1}{4}$ **19.** $\frac{5}{7}$ **20.** −10 **21.** $\frac{-6}{25}$ **22.** $-\frac{21}{40}$ **23.** $\frac{3}{16}$ **24.** $\frac{3}{5}$ **25.** $-1\frac{1}{8}$ **26.** −6

Name ______________________ Date ______________ Class ______________

CHAPTER 2

Family Fun

2A Taskmasters

Directions

- The object of the game is to score the most points.
- Three players play against each other.
- Cut out and shuffle the task cards.
- Player A selects one task card. Player B and Player C complete the task simultaneously.
- If the player is asked to create a problem for the other player, whoever correctly completes the problem first earns the points.
- Player A checks the work. Whoever completes the task/problem correctly earns the points. Answers **MUST** be written in simplest form.
- The players rotate jobs.
- Each player must read the task cards 3 times.
- The player with the most points is the winner!

Task Cards

Create an addition problem for your opponent that involves adding two fractions with unlike denominators. 5 pts	Multiply. Write the answer in simplest form. $\frac{2}{3}\left(\frac{8}{9}\right)$ 2 pts
Create a division problem for your opponent that involves two fractions and whose solution will be negative. 5 pts	Evaluate $4\frac{4}{5}x$ when $x = 20$. 3 pts
Multiply $-9.06\ (0.4)$ 2 pts	Create a multiplication problem that involves a mixed number and a fraction. 5 pts
Solve. Write the answer in simplest form. $x + \frac{7}{8} = \frac{3}{4}$ 3 pts	Solve. Write the answer in simplest form. $\frac{8}{15}t = \frac{7}{9}$ 3 pts
A recipe calls for $\frac{3}{4}$ cups of flour. You need to triple this recipe. How much flour do you need? 2 pts	Subtract. Write the answer in simplest form. $3\frac{7}{12} - 5\frac{1}{9}$ 3 pts

Name ______________________ Date ______________ Class ______________

LESSON 2-1

Practice A

Rational Numbers

Simplify.

1. $\frac{4}{12}$ ______
2. $\frac{5}{15}$ ______
3. $-\frac{2}{8}$ ______
4. $\frac{6}{24}$ ______
5. $\frac{14}{24}$ ______
6. $-\frac{15}{35}$ ______
7. $\frac{10}{21}$ ______
8. $-\frac{16}{36}$ ______

Write each decimal as a fraction in simplest form.

9. 0.4 ______
10. –0.35 ______
11. 0.105 ______
12. 1.2 ______
13. –0.85 ______
14. 0.325 ______
15. 0.002 ______
16. 2.3 ______
17. 0.28 ______
18. –1.25 ______
19. 0.064 ______
20. 0.0075 ______

Write each fraction as a decimal.

21. $\frac{1}{9}$ ______
22. $\frac{9}{16}$ ______
23. $-\frac{11}{20}$ ______
24. $\frac{6}{5}$ ______
25. $\frac{2}{15}$ ______
26. $-2\frac{7}{12}$ ______
27. $\frac{3}{100}$ ______
28. $5\frac{8}{25}$ ______

29. Make up a fraction that cannot be simplified that has 12 as its denominator.

____

Name ______________________ Date ______________ Class ______________

LESSON 2-1

Practice B

Rational Numbers

Simplify.

1. $\frac{6}{9}$ ______
2. $\frac{48}{96}$ ______
3. $\frac{13}{52}$ ______
4. $-\frac{7}{28}$ ______
5. $\frac{15}{40}$ ______
6. $-\frac{4}{48}$ ______
7. $-\frac{14}{63}$ ______
8. $\frac{12}{72}$ ______

Write each decimal as a fraction in simplest form.

9. 0.72 ______
10. 0.058 ______
11. –1.65 ______
12. 2.1 ______
13. 0.036 ______
14. –4.06 ______
15. 2.305 ______
16. 0.0064 ______
17. –0.60 ______
18. 6.95 ______
19. 0.016 ______
20. 0.0005 ______

Write each fraction as a decimal.

21. $\frac{1}{8}$ ______
22. $\frac{8}{3}$ ______
23. $\frac{14}{15}$ ______
24. $\frac{16}{5}$ ______
25. $\frac{11}{16}$ ______
26. $\frac{7}{9}$ ______
27. $\frac{4}{5}$ ______
28. $\frac{31}{25}$ ______

29. Make up a fraction that cannot be simplified that has 24 as its denominator.

____

Name ____________________ Date ______________ Class ______________

LESSON 2-1

Practice C

Rational Numbers

Write each decimal as a fraction in simplest form.

1. 0.9 ________
2. 2.5 ________
3. –0.36 ________
4. –0.215 ________
5. –4.02 ________
6. 0.0085 ________
7. 1.006 ________
8. 0.45 ________

Write each fraction as a decimal.

9. $\frac{9}{15}$ ________
10. $-\frac{22}{50}$ ________
11. $\frac{45}{16}$ ________
12. $-\frac{18}{90}$ ________
13. $\frac{15}{80}$ ________
14. $\frac{21}{126}$ ________
15. $\frac{19}{12}$ ________
16. $\frac{39}{20}$ ________

17. Make up a fraction that cannot be simplified that has 48 as its denominator.

____

18. a. Simplify each fraction below.
    b. Write the denominator of each simplified fraction as the product of prime factors.
    c. Write each fraction as a decimal. Label each as a terminating or repeating decimal.

$\frac{4}{36}$	$\frac{5}{40}$	$\frac{10}{25}$
a. ________	a. ________	a. ________
b. ________	b. ________	b. ________
c. ________	c. ________	c. ________

Name ______________________ Date ______________ Class ______________

LESSON 2-1

Review for Mastery

Rational Numbers

A **rational number** is a *ratio* of two integers.

Rational Number = $\frac{\text{Integer}}{\text{Integer}}$ ← Numerator ← Denominator

Rational Numbers	
Integers	Fractions
Repeating Decimals	Terminating Decimals

The set of rational numbers contains:
- all integers
- all fractions
- decimals that repeat, such as $0.4\overline{6}$
- decimals that terminate, such as 3.5

To simplify a fraction, divide numerator and denominator by the highest common factor.

$$\frac{5}{15} = \frac{5 \div 5}{15 \div 5} = \frac{1}{3}$$

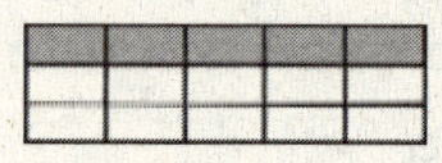

Complete to simplify each fraction.

1. $\frac{8}{16} = \frac{8 \div 8}{16 \div 8} =$ ______

2. $\frac{15}{45} = \frac{\quad \div \quad}{\quad \div \quad} =$ ______

3. $\frac{12}{30} = \frac{\quad \div \quad}{\quad \div \quad} =$ ______

4. $\frac{12}{24} = \frac{\quad \div \quad}{\quad \div \quad} =$ ______

5. $\frac{5}{35} = \frac{\quad \div \quad}{\quad \div \quad} =$ ______

6. $\frac{14}{49} = \frac{\quad \div \quad}{\quad \div \quad} =$ ______

Simplify each fraction.

7. $\frac{8}{56} =$ ______

8. $\frac{15}{50} =$ ______

9. $\frac{8}{36} =$ ______

To write a decimal as a fraction, use the number of decimal places to get the denominator. Then simplify.

$$0.4 = \frac{4}{10} = \frac{4 \div 2}{10 \div 2} = \frac{2}{5}$$

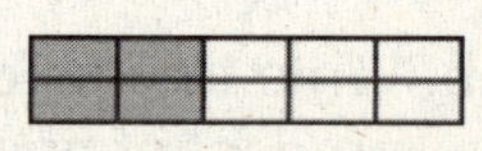

Complete to write each decimal as a fraction in simplest form.

10. $0.25 = \frac{25}{100} = \frac{25 \div \quad}{100 \div \quad} =$ ______

11. $0.375 = \frac{375}{1000} = \frac{375 \div \quad}{1000 \div \quad} =$ ______

Write each decimal as a fraction in simplest form.

12. 0.55 = ______

13. 0.32 = ______

Name ______________________ Date ______________ Class ______________

LESSON 2-1

Review for Mastery

Rational Numbers (continued)

To write a fraction as a decimal, divide numerator by denominator.

A decimal may terminate.

$$\frac{3}{4} = 4\overline{)3.00} = 0.75$$

$$\begin{array}{r} 0.75 \\ 4\overline{)3.00} \\ -28\downarrow \\ \hline 20 \\ -20 \\ \hline 0 \end{array}$$

A decimal may repeat.

$$\frac{1}{3} = 3\overline{)1.00} = 0.\overline{3}$$

$$\begin{array}{r} 0.\overline{3} \\ 3\overline{)1.00} \\ -9\downarrow \\ \hline 10 \\ -9 \\ \hline 1 \end{array}$$

Complete to write each fraction as a decimal.

14. $\frac{15}{4} = 4\overline{)15.00}$

15. $\frac{5}{6} = 6\overline{)5.00}$

16. $\frac{11}{3} = 3\overline{)11.00}$

Write each fraction as a decimal.

17. $\frac{5}{2}$ = ______________

18. $\frac{15}{8}$ = ______________

19. $\frac{28}{6}$ = ______________

20. $\frac{22}{4}$ = ______________

21. $\frac{62}{12}$ = ______________

22. $\frac{105}{10}$ = ______________

Name ____________________ Date ____________ Class ____________

LESSON 2-1

Challenge

Encore, Encore, ...

Explore some patterns with repeating decimals. Use a calculator to write each decimal equivalent.

1. $\frac{1}{9}$ = ____________ 2. $\frac{2}{9}$ = ____________ 3. $\frac{3}{9}$ = ____________

Predict the decimal equivalent of each fraction. Verify your results on a calculator.

4. $\frac{4}{9}$ = ____________ 5. $\frac{6}{9}$ = ____________ 6. $\frac{8}{9}$ = ____________

Write each fractional equivalent.

7. $0.\overline{5}$ = ____________ 8. $0.\overline{7}$ = ____________ 9. $0.\overline{9}$ = ____________

Use a calculator to write each decimal equivalent.

10. $\frac{42}{99}$ = ____________ 11. $\frac{358}{999}$ = ____________ 12. $\frac{4276}{9999}$ = ____________

Predict the decimal equivalent of each fraction.

13. $\frac{76}{99}$ = ____________ 14. $\frac{732}{999}$ = ____________ 15. $\frac{1957}{9999}$ = ____________

Write each fractional equivalent.

16. $0.\overline{45}$ = ____________ 17. $0.\overline{148}$ = ____________ 18. $0.\overline{7213}$ = ____________

19. Summarize your observations.

____

____

____

Name ______________________ Date ______________ Class ______________

LESSON 2-1

Problem Solving

Rational Numbers

Write the correct answer.

1. Fill in the table below which shows the sizes of drill bits in a set.

2. Do the drill bit sizes convert to repeating or terminating decimals?

____

13-Piece Drill Bit Set

Fraction	Decimal	Fraction	Decimal	Fraction	Decimal
$\frac{1}{4}$"		$\frac{11}{64}$"		$\frac{3}{32}$"	
$\frac{15}{64}$"		$\frac{5}{32}$"		$\frac{5}{64}$"	
$\frac{7}{32}$"		$\frac{9}{64}$"		$\frac{1}{16}$"	
$\frac{13}{64}$"		$\frac{1}{8}$"			
$\frac{3}{16}$"		$\frac{7}{64}$"			

Use the table at the right that lists the world's smallest nations. Choose the letter for the best answer.

Three of the World's Smallest Nations

Nation	**Area** (square miles)
Vatican City	0.17
Monaco	0.75
Nauru	8.2

3. What is the area of Vatican City expressed as a fraction in simplest form?

A $\frac{8}{50}$ C $\frac{17}{1000}$

B $\frac{4}{25}$ D $\frac{17}{100}$

4. What is the area of Monaco expressed as a fraction in simplest form?

F $\frac{75}{100}$ H $\frac{3}{4}$

G $\frac{15}{20}$ J $\frac{2}{3}$

5. What is the area of Nauru expressed as a mixed number?

A $8\frac{1}{50}$ C $8\frac{2}{100}$

B $8\frac{2}{50}$ D $8\frac{1}{5}$

6. The average annual precipitation in Miami, FL is 57.55 inches. Express 57.55 as a mixed number.

F $57\frac{11}{20}$ H $57\frac{5}{100}$

G $57\frac{55}{1000}$ J $57\frac{1}{20}$

7. The average annual precipitation in Norfolk, VA is 45.22 inches. Express 45.22 as a mixed number.

A $45\frac{11}{50}$ C $45\frac{11}{20}$

B $45\frac{22}{1000}$ D $45\frac{1}{5}$

Name ______________________ Date ______________ Class ______________

LESSON 2-1

Reading Strategies

Use a Graphic Organizer

Definition	Facts
The set of numbers that can be written in the form $\frac{a}{b}$, where *a* and *b* are integers and *b* does not equal 0.	Fractions are rational numbers. Decimals that terminate or repeat are rational numbers. Whole numbers are rational numbers. Integers are rational numbers. 0 is a rational number.
Examples	**Non-examples**
$2 = \frac{2}{1}$ $\frac{7}{8}$ $0.37 = \frac{37}{100}$ $4\frac{1}{4} = \frac{17}{4}$ $-5 = -\frac{5}{1}$	$\sqrt{3}$ (the square root of 3) π (3.14159…) $\sqrt{3}$ and π cannot be written as decimals that terminate or repeat.

Rational Numbers

Use the chart to answer the following questions.

1. What is a rational number?

____

2. Is 0.62 a rational number? Why or why not?

____

3. Is $2\frac{1}{3}$ a rational number? Why or why not?

____

4. Is $\sqrt{7}$ a rational number? Why or why not?

____

5. Is –8 a rational number? Why or why not?

____

6. Is 0 a rational number? Why or why not?

____

Name ______________________ Date ______________ Class ______________

LESSON 2-1

Puzzles, Twisters & Teasers

Let's Be Rational!

Circle words from the list in the word search. Then find a word that answers the riddle. Circle it and write it on the line.

rational	equivalent	numerator	denominator	relatively
prime	simplify	integer	nonzero	factor

```
H O L E Q U I V A L E N T N C
M I R U J I N O N Z E R O U Y
O P R F A C T O R X D E B M F
Q W A U I O E W E R T Y L E I
O T T L P J G O K I Y T R R L
B Y I M J U E W E R T Y U A P
N J O A S P R I M E T Y U T M
D E N O M I N A T O R I L O I
X W A R E L A T I V E L Y R S
P L L Q A Z X S W E D C V F R
```

What can you put in a bucket of water to make it lighter?

a ___ ___ ___ ___

Name ______________________ Date ______________ Class ______________

LESSON 2-2

Practice A

Comparing and Ordering Rational Numbers

Compare. Write <, >, or =.

1. $\frac{3}{4}$ ■ $\frac{5}{7}$

$\downarrow$ $\downarrow$

$\frac{\_\_}{28}$ ___ $\frac{\_\_}{28}$

$\downarrow$ $\downarrow$

$\frac{3}{4}$ ___ $\frac{5}{7}$

2. $-\frac{2}{5}$ ■ $-\frac{3}{8}$

$-\frac{\_\_}{40}$ ___ $-\frac{\_\_}{40}$, so

$-\frac{2}{5}$ ___ $-\frac{3}{8}$

3. 0.3 ■ $\frac{1}{4}$

0.3 ___ ____

0.3 ___ $\frac{1}{4}$

4. $\frac{1}{6}$ ■ $\frac{1}{3}$

$\downarrow$ $\downarrow$

___ ___ ___

$\frac{1}{6}$ ___ $\frac{1}{3}$

5. 0.09 ■ $\frac{1}{2}$

$\downarrow$ $\downarrow$

___ ___ ___

0.09 ___ $\frac{1}{2}$

6. $-\frac{3}{5}$ ■ –0.6

$\downarrow$ $\downarrow$

___ ___ ___

$-\frac{3}{5}$ ___ –0.6

7. $\frac{4}{5}$ ___ $\frac{7}{10}$

8. $-\frac{1}{4}$ ___ $-\frac{3}{4}$

9. $-\frac{2}{3}$ ___ $-\frac{1}{4}$

10. $\frac{5}{8}$ ___ $\frac{5}{6}$

11. $\frac{7}{9}$ ___ $\frac{2}{3}$

12. $-\frac{3}{5}$ _____ $-\frac{1}{10}$

13. $\frac{0}{1}$ ___ $\frac{1}{8}$

14. $-1\frac{4}{5}$ ___ –1.3

15. $1\frac{2}{3}$ ___ $1\frac{2}{5}$

16. Trail A is $2\frac{2}{5}$ miles long. Trail B is $\frac{1}{4}$ mile long. Trail C is $1\frac{9}{10}$ miles long. Trail D is 2.05 miles long. List the lengths of the trails from shortest to longest.

____

Name ______________________ Date ______________ Class ______________

LESSON 2-2

Practice B

Comparing and Ordering Rational Numbers

Compare. Write <, >, or =.

1. $\frac{1}{8}$ ___ $\frac{1}{10}$

2. $\frac{3}{5}$ ___ $\frac{7}{10}$

3. $-\frac{1}{3}$ ___ $-\frac{3}{4}$

4. $\frac{5}{6}$ ___ $\frac{3}{4}$

5. $-\frac{2}{7}$ ___ $-\frac{1}{2}$

6. $1\frac{2}{9}$ ___ $1\frac{2}{3}$

7. $-\frac{8}{9}$ ___ $-\frac{3}{10}$

8. $-\frac{4}{5}$ ___ $-\frac{8}{10}$

9. 0.08 ___ $\frac{3}{10}$

10. $\frac{11}{15}$ ___ $0.7\overline{3}$

11. $2\frac{4}{9}$ ___ $2\frac{3}{4}$

12. $-\frac{5}{8}$ ___ -0.58

13. $3\frac{1}{4}$ ___ 3.3

14. $-\frac{1}{6}$ ___ $-\frac{1}{9}$

15. 0.75 ___ $\frac{3}{4}$

16. $-2\frac{1}{8}$ ___ -2.1

17. $1\frac{1}{2}$ ___ 1.456

18. $-\frac{3}{5}$ ___ -0.6

19. On Monday, Gina ran 1 mile in 9.3 minutes. Her times for running 1 mile on each of the next four days, relative to her time on Monday, were $-1\frac{2}{3}$ minutes, -1.45 minutes, -1.8 minutes, and $-1\frac{3}{8}$ minutes. List these relative times in order from least to greatest.

____

20. Trail A is 3.1 miles long. Trail C is $3\frac{1}{4}$ miles long. Trail B is longer than Trail A but shorter than Trail C. What is a reasonable distance for the length of Trail B?

____

Name ______________________ Date ______________ Class ______________

LESSON 2-2

Practice C

Comparing and Ordering Rational Numbers

Compare. Write <, >, or =.

1. $\frac{1}{9}$ ___ $\frac{1}{20}$
2. $\frac{3}{8}$ ___ $\frac{5}{6}$
3. $-\frac{1}{6}$ ___ $-\frac{2}{3}$
4. $\frac{11}{20}$ ___ 0.55
5. $-\frac{4}{9}$ ___ $-\frac{1}{2}$
6. $1\frac{3}{5}$ ___ 1.35
7. $-\frac{5}{9}$ ___ −0.45
8. $-1\frac{7}{8}$ ___ −1.875
9. $\frac{5}{4}$ ___ 1.4
10. $-1\frac{1}{5}$ ___ −1.06
11. 4.00 ___ $\frac{24}{4}$
12. $-\frac{5}{12}$ ___ −0.56

Write a fraction or decimal that has a value between the given numbers.

13. $\frac{1}{6}$ and $\frac{1}{5}$ ______
14. 0.7 and 0.71 ______
15. $-\frac{1}{6}$ and 0.2 ______
16. $-\frac{1}{2}$ and $-\frac{1}{4}$ ______
17. 1.45 and 1.46 ______
18. $\frac{2}{3}$ and 0.75 ______

19. The students in one English class are reading the same book. Last night, James read $\frac{1}{4}$ of the book. Jennie read $\frac{3}{8}$ of the book. Kyle read 0.4 of the book, and Talia read 0.33 of the book. List the numbers in order from least to greatest. Who read the greatest number of pages last night?

____

20. Melanie ran $2\frac{5}{8}$ miles on Monday. On Friday she ran 2.8 miles. On Wednesday, she ran further than on Monday but not as far as on Friday. What is a reasonable fraction length for the distance Melanie ran on Wednesday? What is a reasonable decimal length for the distance?

____

Name ______________________ Date ______________ Class ______________

LESSON 2-2

Review for Mastery

Comparing and Ordering Rational Numbers

You can use number lines to compare two fractions that have different denominators.

Compare $\frac{3}{8}$ and $\frac{2}{3}$.

Compare $-\frac{3}{4}$ and $-\frac{5}{6}$.

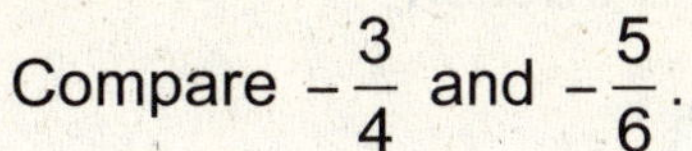

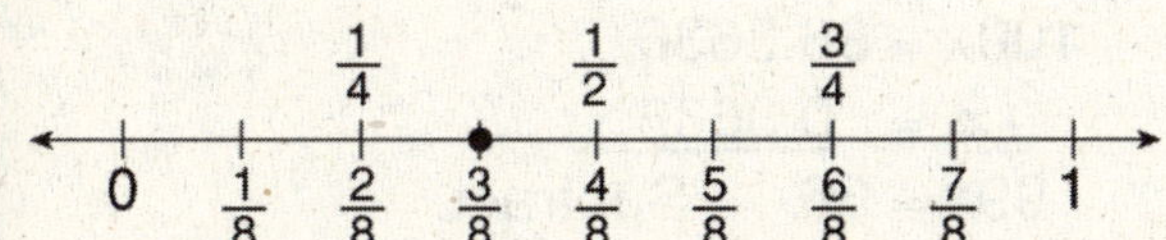

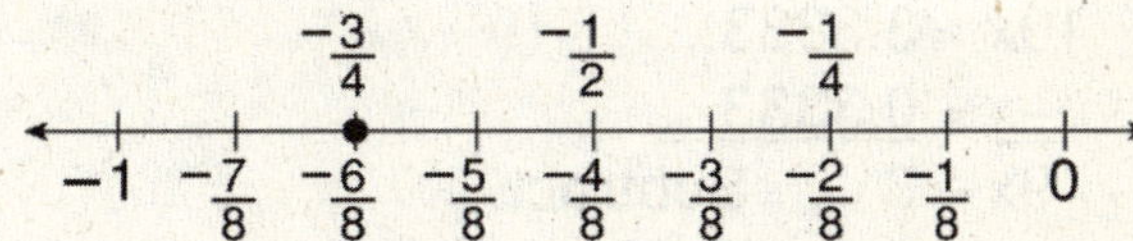

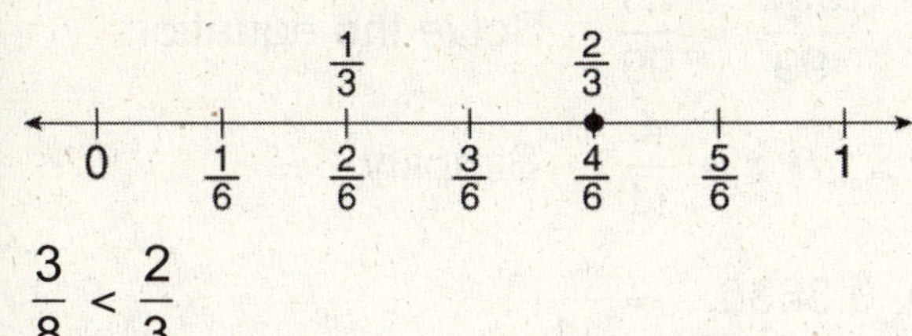

$\frac{3}{8} < \frac{2}{3}$

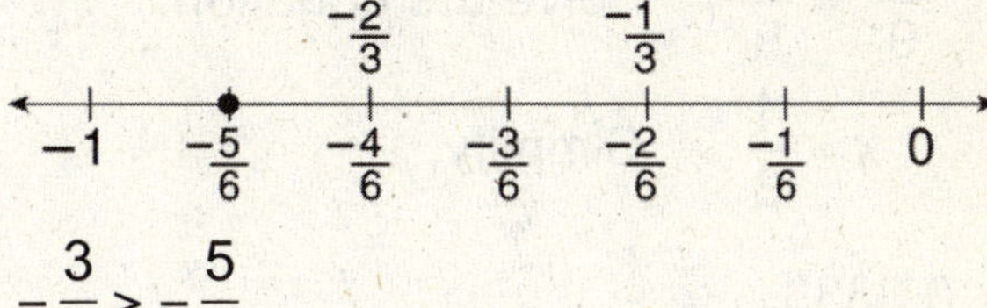

$-\frac{3}{4} > -\frac{5}{6}$

Use the number lines above. Write < or >.

1. $\frac{1}{4}$ ___ $\frac{1}{6}$
2. $\frac{5}{6}$ ___ $\frac{1}{2}$
3. $-\frac{2}{3}$ ___ $-\frac{1}{4}$
4. $-\frac{5}{6}$ ___ $-\frac{5}{8}$

You can also use number lines to compare a fraction and a decimal.

Compare 0.2 and $\frac{1}{3}$.

Compare $-\frac{5}{8}$ and –0.9.

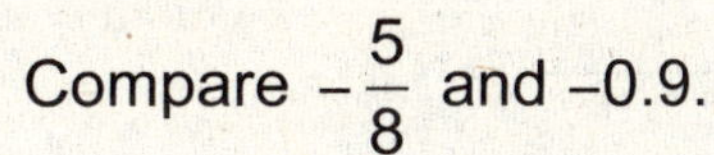

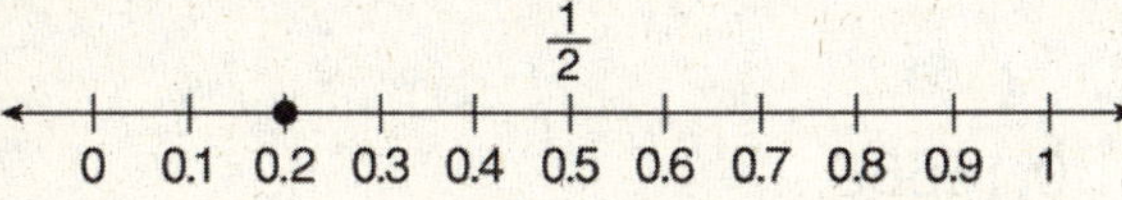

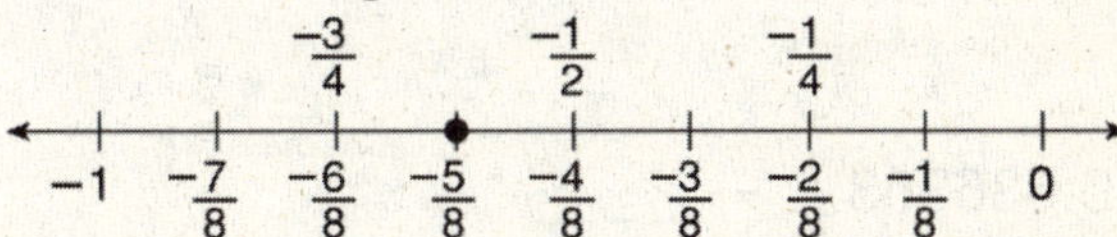

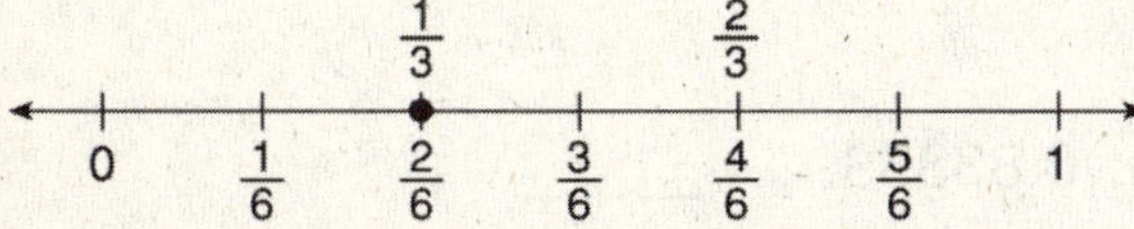

$0.2 < \frac{1}{3}$

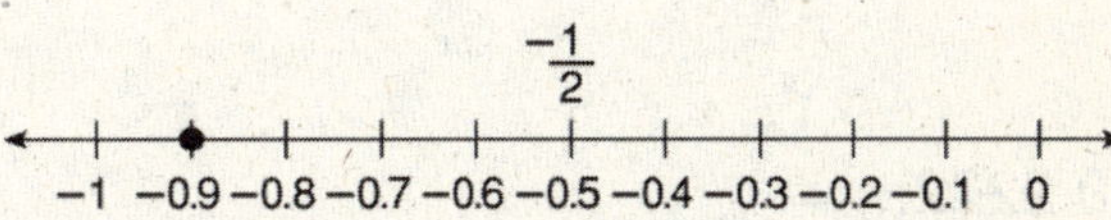

$-\frac{5}{8} > -0.9$

Use the number lines above. Write < or >.

5. $\frac{5}{6}$ ___ 0.5
6. 0.6 ___ $\frac{2}{3}$
7. –0.4 ___ $-\frac{1}{4}$
8. $-\frac{7}{8}$ ___ –0.8

Name ______________________ Date ______________ Class ______________

LESSON 2-2

Challenge

From Repeating Decimal to Fraction

You can use an equation to write a repeating decimal as a fraction.

Write 0.3333… as a fraction.	Write 0.363636… as a fraction.
Let $x = 0.3333...$ Then $10x = 3.3333...$ $10x = 3.3333...$ $\underline{-x = 0.3333...}$ $9x = 3$ Subtract. $\frac{9x}{9} = \frac{3}{9}$ Solve the equation. $x = \frac{1}{3}$ Simplify. So, $0.3333... = \frac{1}{3}$.	Let $x = 0.363636...$ Then $100x = 36.3636336...$ $100x = 36.3636...$ $\underline{-x = 0.3636...}$ $99x = 36$ Subtract. $\frac{99x}{99} = \frac{36}{99}$ Solve the equation. $x = \frac{4}{11}$ Simplify. So, $0.3636... = \frac{4}{11}$.

Write each repeating decimal as a fraction.

1. 0.6666… = ______
2. 0.8888… = ______
3. 0.454545… = ______
4. 0.090909… = ______
5. 0.636363… = ______
6. 0.16666… = ______
7. 0.2222… = ______
8. 0.83333… = ______
9. 0.41666… = ______
10. 0.58333… = ______

Name ______________________ Date ______________ Class ______________

LESSON 2-2

Problem Solving

Comparing and Ordering Rational Numbers

Write the correct answer.

1. Carl Lewis won the gold medal in the long jump in four consecutive Summer Olympic games. He jumped 8.54 meters in 1984, 8.72 meters in 1988, 8.67 meters in 1992, 8.5 meters in 1996. Order the length of his winning jumps from least to greatest.

____

2. The depth of a lake is measured at three different points. Point A is −15.8 meters, Point B is −17.3 meters, and Point C is −16.9 meters. Which point has the greatest depth?

____

3. Scientists aboard a submarine are gathering data at an elevation of $-42\frac{1}{2}$ feet. Scientists aboard a submersible are taking photographs at an elevation of $-45\frac{1}{3}$ feet. Which scientists are closer to the surface of the ocean?

____

4. At a swimming meet, Gail's time in her first heat was $42\frac{3}{8}$ seconds. Her time in the second heat was 42.25 seconds. Which heat did she swim faster?

____

The table shows the top times in a 5 K race. Choose the letter of the best answer.

Name	Time (minutes)
Marshall	18.09
Renzo	17.38
Dan	17.9
Aaron	18.61

5. Who had the fastest time in the race?
   A Marshall
   B Renzo
   C Dan
   D Aaron

6. Which is the slowest time in the table?
   F 18.09 minutes
   G 17.38
   H 17.9 minutes
   J 18.61 minutes

7. Aaron's time in a previous race was less than his time in this race but greater than Marshall's time in this race. How fast could Aaron have run in the previous race?
   A 19.24 min
   B 18.7 min
   C 18.35 min
   D 18.05 mi

Name ______________________ Date ____________ Class ____________

LESSON 2-2

Reading Strategies

Identify Relationships

0, $\frac{1}{2}$, and 1 are common benchmarks for fractions. Sometimes you can use these benchmarks to compare fractions.

A fraction is close to 0 if its numerator is small compared to its denominator. Examples are $\frac{2}{11}$, $\frac{3}{20}$, and $\frac{4}{25}$.

A fraction is close to $\frac{1}{2}$ if its denominator is about twice as great as its denominator. Examples are $\frac{5}{11}$, $\frac{8}{15}$, and $\frac{10}{21}$.

A fraction is close to 1 if its numerator and denominator are close in value. Examples are $\frac{9}{10}$, $\frac{13}{15}$, and $\frac{17}{33}$.

To compare $\frac{7}{15}$ and $\frac{3}{22}$, look at the relationship between the numerator and denominator of each fraction.

15 is a little more than 2×7, so $\frac{1}{2}$ is a benchmark for $\frac{7}{15}$.

3 is much smaller than 22, so 0 is a benchmark for $\frac{3}{22}$.

Since $\frac{1}{2} > 0$, $\frac{7}{15} > \frac{3}{22}$.

Answer each question.

1. What is a benchmark for $\frac{35}{67}$? ____________
2. What is a benchmark for $\frac{11}{13}$? ____________
3. Use > or < to compare $\frac{35}{67}$ and $\frac{11}{13}$. ____________
4. What is a benchmark for $\frac{17}{20}$? ____________
5. What is a benchmark for $\frac{6}{35}$? ____________
6. Use > or < to compare $\frac{17}{20}$ and $\frac{6}{35}$. ____________
7. Use benchmarks to compare $\frac{21}{40}$ and $\frac{19}{21}$. Explain your thinking.

____

Name ______________________ Date ______________ Class ______________

LESSON 2-2

Puzzles, Twisters & Teasers

Rational Riddle

Why did Francis Fraction want to become a psychologist?

To answer the riddle, write the numbers in order from least to greatest. Then write the corresponding letters in the same order.

$-1\frac{1}{4}$	**A**	1.	________	________
-2.05	**E**	2.	________	________
$-\frac{1}{5}$	**T**	3.	________	________
$\frac{3}{4}$	**N**	4.	________	________
0	**I**	5.	________	________
-2.7	**S**	6.	________	________
$-\frac{5}{6}$	**R**	7.	________	________
$-1\frac{1}{8}$	**S**	8.	________	________
0.95	**L**	9.	________	________
-0.6	**A**	10.	________	________
-1.3	**W**	11.	________	________
0.8	**A**	12.	________	________
$-2\frac{1}{2}$	**H**	13.	________	________
$\frac{1}{8}$	**O**	14.	________	________

Answer: ______________________________

Name ______________________ Date ______________ Class ______________

LESSON 2-3

Practice A
Adding and Subtracting Rational Numbers

1. A statue $8\frac{5}{16}$ in. high rests on a stand that is $1\frac{3}{16}$ in. high. What is the total height?

____

2. During the 19th Olympic Winter Games in 2002, the United States 4-man bobsled teams won silver and bronze medals. USA-1 sled had a total time of 3 min 7.81 sec. The USA-2 sled had a total time of 3 min 7.86 sec. What is the difference in the time of the two runs?

____

Use a number line to find each sum.

3. $-0.2 + 0.6$

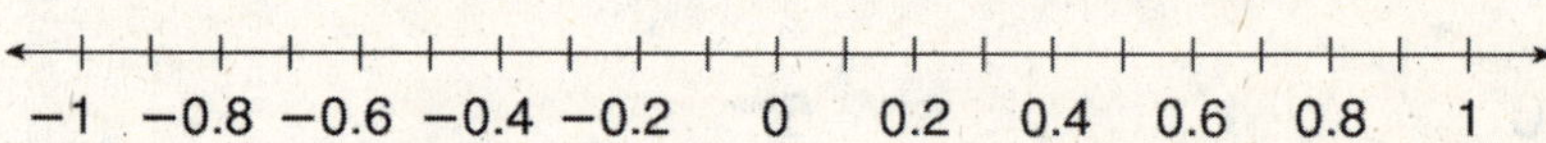

4. $\frac{1}{5} + \frac{3}{5}$

____

$-1 \quad -\frac{4}{5} \quad -\frac{3}{5} \quad -\frac{2}{5} \quad -\frac{1}{5} \quad 0 \quad \frac{1}{5} \quad \frac{2}{5} \quad \frac{3}{5} \quad \frac{4}{5} \quad 1$

Add or subtract. Write each answer in simplest form.

5. $\frac{2}{9} + \frac{4}{9}$ ____________
6. $\frac{5}{12} + \frac{3}{12}$ ____________
7. $\frac{9}{10} - \frac{7}{10}$ ____________
8. $\frac{8}{15} - \frac{11}{15}$ ____________
9. $\frac{3}{14} - \frac{9}{14}$ ____________
10. $\frac{5}{18} - \frac{11}{18}$ ____________
11. $\frac{1}{8} + \frac{5}{8}$ ____________
12. $\frac{5}{6} + \frac{1}{6}$ ____________

Evaluate each expression for the given value of the variable.

13. $18.3 + x$ for $x = -1.6$ ____________
14. $20.6 + x$ for $x = 2.8$ ____________
15. $\frac{9}{11} + x$ for $x = -\frac{5}{11}$ ____________

Name ______________________ Date ______________ Class ______________

LESSON 2-3

Practice B

Adding and Subtracting Rational Numbers

1. Gretchen bought a sweater for \$23.89. In addition, she had to pay \$1.43 in sales tax. She gave the sales clerk \$30. How much change did Gretchen receive from her total purchase?

____

2. Jacob is replacing the molding around two sides of a picture frame. The measurements of the sides of the frame are $4\frac{3}{16}$ in. and $2\frac{5}{16}$ in. What length of molding will Jacob need?

____

Use a number line to find each sum.

3. $-0.5 + 0.4$

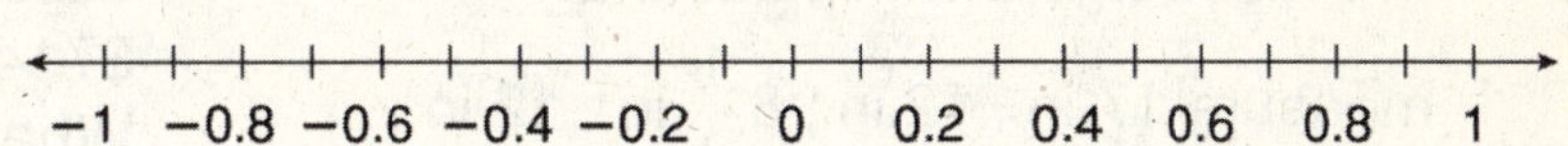

____

4. $-\frac{2}{7} + \frac{6}{7}$

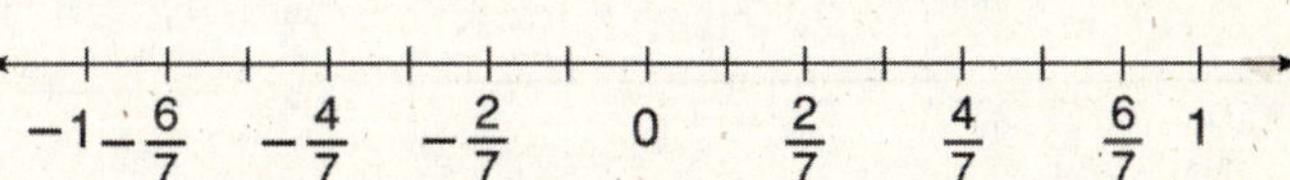

____

Add or subtract. Write each answer in simplest form.

5. $\frac{3}{8} + \frac{1}{8}$ ______________________

6. $-\frac{1}{10} + \frac{7}{10}$ ______________________

7. $\frac{5}{14} - \frac{3}{14}$ ______________________

8. $\frac{4}{15} + \frac{7}{15}$ ______________________

9. $\frac{5}{18} - \frac{7}{18}$ ______________________

10. $-\frac{8}{17} - \frac{2}{17}$ ______________________

11. $-\frac{1}{16} + \frac{5}{16}$ ______________________

12. $\frac{3}{20} + \frac{1}{20}$ ______________________

Evaluate each expression for the given value of the variable.

13. $38.1 + x$ for $x = -6.1$ ______________________

14. $18.7 + x$ for $x = 8.5$ ______________________

15. $\frac{8}{15} + x$ for $x = -\frac{4}{15}$ ______________________

Name ______________________ Date ______________ Class ______________

LESSON 2-3

Practice C

Adding and Subtracting Rational Numbers

1. Jesse baked a pizza and cut it into 8 pieces. He ate three pieces and his two brothers ate two pieces each. In fractional form, how much of the pizza is left?

____

2. The biathlon combines cross-country skiing with rifle shooting at fixed targets. A biathlon competition features races of 6.2 mi with the contestants stopping to shoot twice, a race of 12.45 mi with four shooting stops, and a four-person relay race totaling 18.6 mi. How many total miles are covered in a biathlon competition?

____

3. The home plate in baseball is a five-sided figure with sides that measure 17 in., 12 in., $8\frac{1}{2}$ in., 12 in., and $8\frac{1}{2}$ in. What is the sum of the sides of the figure?

____

4. Chato had \$2032.64 in his checking account. He wrote two checks, one for \$714.53 and the other for \$289.67. What was the new balance in his checking account?

____

Add or subtract. Write each answer in simplest form.

5. $\frac{9}{17} + \frac{3}{17}$ ____________

6. $-\frac{4}{25} + \frac{19}{25}$ ____________

7. $\frac{17}{30} + \frac{7}{30}$ ____________

8. $-\frac{13}{20} + \frac{9}{20}$ ____________

9. $-\frac{8}{15} - \frac{4}{15}$ ____________

10. $\frac{5}{24} + \frac{13}{24}$ ____________

11. $\frac{13}{18} - \frac{7}{18}$ ____________

12. $-\frac{33}{50} - \frac{7}{50}$ ____________

Evaluate each expression for the given value of the variable.

13. $102.943 + x$ for $x = 2.03$

____

14. $\frac{12}{25} - x$ for $x = -\frac{13}{25}$

____

15. $18.01 - x$ for $x = -19.26$

____

Name ______________________ Date ______________ Class ______________

LESSON 2-3

Review for Mastery

Adding and Subtracting Rational Numbers

To add fractions that have the same denominator:

- Use the common denominator for the sum.
- Add the numerators to get the numerator of the sum.
- Write the sum in simplest form.

$$\frac{1}{8}+\frac{3}{8}=\frac{1+3}{8}=\frac{4}{8}=\frac{1}{2}$$

To subtract fractions that have the same denominator:

- Use the common denominator for the difference.
- Subtract the numerators.
  Subtraction is addition of an opposite.
- Write the difference in simplest form.

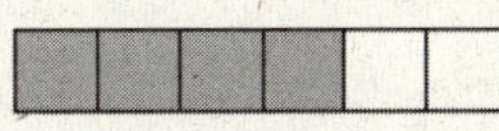

$$\frac{3}{6}-\left(-\frac{1}{6}\right)=\frac{3+1}{6}=\frac{4}{6}=\frac{2}{3}$$

Complete to add the fractions.

1. $\frac{3}{14}+\frac{4}{14}=$ ____ = ____

2. $\frac{2}{10}+\left(-\frac{4}{10}\right)=$ ____ = ____

3. $-\frac{5}{12}+\left(-\frac{3}{12}\right)=$ ____ = ____

Complete to subtract the fractions.

4. $\frac{8}{9}-\frac{2}{9}=$ ____ = ____

5. $\frac{9}{15}-\left(-\frac{3}{15}\right)=$ ____ = ____

6. $-\frac{10}{24}-\left(-\frac{2}{24}\right)=$ ____ = ____

To add or subtract decimals, line up the decimal points and then add or subtract from right to left as usual.

12.83	35.78
+24.17	−14.55
37.00	21.23

Complete to add the decimals.

7. 14.23 + 3.56 = ______________

8. 44.02 + 8.07 = ______________

9. 1.39 + 13.6 = ______________

Complete to subtract the decimals.

10. 124.33 − 13.16 = ______________

11. 33.47 − 0.6 = ______________

12. 25.15 − 25.06 = ______________

Name ______________________ Date ______________ Class ______________

LESSON 2-3

Challenge

Number Code

Each sum is the code for a letter. As you find a sum, write its letter code in the message below. Write the sum in simplest form. Some letters appear more than once. An example is done for you.

$4.5 + (-6.5)$ **−2**, T

1. $14.56 + (-10.09)$ ________, V

2. $\frac{7}{8} + \left(-1\frac{3}{8}\right)$ ________, M

3. $\frac{6}{8} + \left(-\frac{3}{8}\right)$ ________, N

4. $-1.05 + 0.85$ ________, I

5. $\frac{-2}{4} + \left(\frac{-3}{4}\right)$ ________, U

6. $-7.08 + (-12.02)$ ________, S

7. $-9.5 + 3.1$ ________, E

8. $\frac{-4}{5} + 1$ ________, E

9. $-1\frac{1}{2} + \left(-1\frac{1}{2}\right)$ ________, H

10. $1\frac{2}{4} + \left(\frac{-3}{4}\right)$ ________, P

11. $5 + \left(-4\frac{1}{10}\right)$ ________, I

12. $8 + (-6.4)$ ________, Y

13. $-3\frac{1}{4} + 3\frac{1}{4}$ ________, S

14. $\frac{7}{8} + \left(-1\frac{7}{8}\right)$ ________, L

15. $6.52 + (-5)$ ________, Z

16. $-62.3 + 23.9$ ________, A

17. $9\frac{1}{8} + (-10)$ ________, R

18. $-2.9 + 0.85$ ________, O

19. $2.7 + (-0.9)$ ________, O

___ ___ ___ ___   ___ ___ ___   ___ ___ ___

-19.1 $-1\frac{1}{4}$ $-\frac{7}{8}$ -6.4 1.6 -2.05 $-1\frac{1}{4}$ -38.4 $-\frac{7}{8}$ $\frac{1}{5}$

___ ___ **T**   ___ ___ ___ ___ **T** ___ ___ ___

$\frac{3}{8}$ 1.8 -2 -1 $\frac{1}{5}$ -19.1 0 -2 -3 -38.4 $\frac{3}{8}$

___ ___ ___ ___?

1.52 -6.4 $-\frac{7}{8}$ -2.05

___ ___   ___ ___ ___ ___   **T** ___ ___ ___!

-0.2 $-\frac{1}{2}$ $\frac{3}{4}$ -2.05 0 $\frac{9}{10}$ -2 -0.2 4.47 $\frac{1}{5}$

Name ______________________ Date ____________ Class ____________

LESSON 2-3

Problem Solving

Adding and Subtracting Rational Numbers

Write the correct answer.

1. At a track meet, a sprinter won the 100-m dash with a time of 10.93 seconds. At another meet, a second sprinter won the 100-m dash with a time of 10.75 seconds. How many seconds faster did the second sprinter run the 100-m dash?

____

2. The snowfall in Rochester, NY in the winter of 1999–2000 was 91.5 inches. Normal snowfall is about 76 inches per winter. How much more snow fell in the winter of 1999–2000 than is normal?

____

3. In a survey, $\frac{76}{100}$ people indicated that they check their e-mail daily, while $\frac{23}{100}$ check their e-mail weekly, and $\frac{1}{100}$ check their e-mail less than once a week. What fraction of people check their e-mail at least once a week?

____

4. To make a small amount of play dough, you can mix the following ingredients: 1 cup of flour, $\frac{1}{2}$ cup of salt and $\frac{1}{2}$ cup of water. What is the total amount of ingredients added to make the play dough?

____

Choose the letter for the best answer.

Baseball Ticket Prices

Location	Average Price
Minnesota	$14.42
League Average	$19.82
Boston	$40.77

5. How much more expensive is it to buy a ticket in Boston than in Minnesota?

A $20.95
B $55.19
C $5.40
D $26.35

6. How much more expensive is it to buy a ticket in Boston than the league average?

F $60.59
G $20.95
H $5.40
J $26.35

7. What is the total cost of a ticket in Boston and a ticket in Minnesota?

A $55.19
B $34.24
C $60.59
D $54.19

Name ______________________ Date ______________ Class ______________

LESSON 2-3

Reading Strategies

Use a Visual Model

A number line can help you picture addition with decimals. This number line is divided into tenths.

Add – 0.6 + 1.8.

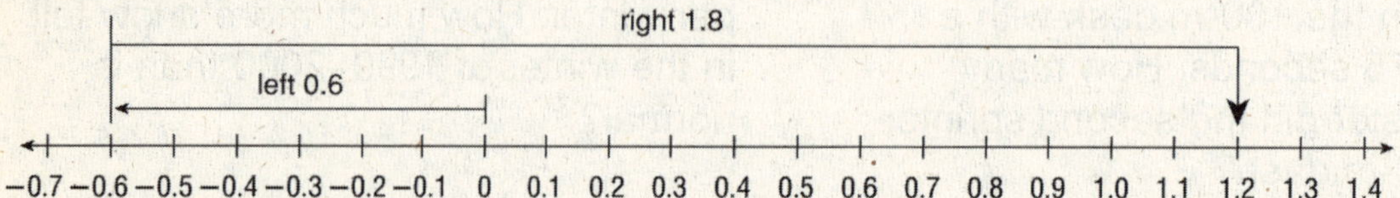

1. Where do you start on the number line? ______________
2. Do you move to the right or to the left? Why?

____

3. How many places do you move? ______________
4. To add, do you move to the right or to the left? ______________
5. How many places do you move? ______________
6. At what number do you end? ______________

This number line helps you picture addition with fractions. The number line is divided into sixths.

Add $-\frac{4}{6} + 1\frac{3}{6}$.

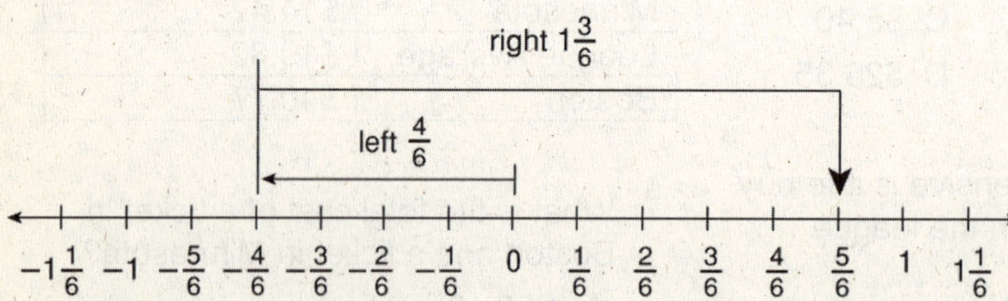

7. Do you move first to the left or right from 0 on the number line? Why?

____

8. How many places do you move? ______________
9. To add, do you move to the right or left? ______________
10. How many places do you move? ______________
11. At what number do you end? ______________

Name ______________________ Date ______________ Class ______________

LESSON 2-3

Puzzles, Twisters & Teasers

Bee a Math Master!

Add or subtract to find the answers. Then solve the riddle using the letters associated with the answers.

S $-\frac{1}{12} + (-\frac{7}{12}) =$ ______________

F $-0.9 + 2.5 =$ ______________

D $\frac{8}{11} - \frac{3}{11} =$ ______________

O $-\frac{4}{13} - \frac{8}{13} =$ ______________

R $-\frac{1}{15} + \frac{13}{15} =$ ______________

G $\frac{11}{32} - \frac{27}{32} =$ ______________

W $-0.06 + 0.86 =$ ______________

T $-\frac{19}{25} + \frac{13}{25} =$ ______________

E $\frac{8}{21} + \frac{15}{21} =$ ______________

H $0.9 + 0.3 =$ ______________

Why did the bee hum?

It ___ ___ ___ ___ ___ ___   ___ ___ ___

1.6 $-\frac{12}{13}$ $\frac{4}{5}$ $-\frac{1}{2}$ $-\frac{12}{13}$ $-\frac{6}{25}$ $-\frac{6}{25}$ 1.2 $\frac{23}{21}$

___ ___ ___ ___ ___

0.8 $-\frac{12}{13}$ $\frac{4}{5}$ $\frac{5}{11}$ $-\frac{2}{3}$

Name ______________________ Date ______________ Class ______________

LESSON 2-4

Practice A

Multiplying Rational Numbers

Multiply. Write each answer in simplest form.

1. $5\left(\frac{1}{3}\right)$

2. $-2\left(\frac{2}{5}\right)$

3. $4\left(\frac{1}{6}\right)$

4. $-3\left(\frac{2}{9}\right)$

5. $-\frac{5}{7}\left(\frac{2}{5}\right)$

6. $\frac{3}{4}\left(\frac{1}{3}\right)$

7. $-\frac{1}{4}\left(\frac{1}{3}\right)$

8. $-\frac{1}{6}\left(-\frac{2}{3}\right)$

9. $\frac{1}{2}\left(\frac{10}{7}\right)$

10. $\frac{3}{10}\left(-\frac{5}{18}\right)$

11. $\frac{4}{5}\left(-\frac{12}{16}\right)$

12. $\frac{4}{3}\left(\frac{24}{16}\right)$

13. $-4\left(1\frac{1}{2}\right)$

14. $\frac{3}{4}\left(\frac{5}{8}\right)$

15. $-\frac{2}{5}\left(3\frac{1}{4}\right)$

16. $-\frac{5}{6}\left(-\frac{3}{10}\right)$

Multiply.

17. -3.2×5

18. 0.34×0.06

19. -8.12×-9

20. 4.24×3.5

21. -3.14×0.007

22. -6.7×0.8

23. -0.25×-2.4

24. 7.9×-2

25. Jade babysat $4\frac{1}{2}$ hours for the Lenox family. She was paid \$5 an hour. How much did she receive for this babysitting job?

Name ______________________ Date ____________ Class ____________

LESSON 2-4

Practice B

Multiplying Rational Numbers

Multiply. Write each answer in simplest form.

1. $8\left(\frac{3}{4}\right)$
2. $-6\left(\frac{9}{18}\right)$
3. $-9\left(\frac{5}{6}\right)$
4. $-6\left(-\frac{7}{12}\right)$
5. $-\frac{5}{18}\left(\frac{8}{15}\right)$
6. $\frac{7}{12}\left(\frac{14}{21}\right)$
7. $-\frac{1}{9}\left(\frac{27}{24}\right)$
8. $-\frac{1}{11}\left(-\frac{3}{2}\right)$
9. $\frac{7}{20}\left(-\frac{15}{28}\right)$
10. $\frac{16}{25}\left(-\frac{18}{32}\right)$
11. $\frac{1}{9}\left(-\frac{18}{17}\right)$
12. $\frac{17}{20}\left(-\frac{12}{34}\right)$
13. $-4\left(2\frac{1}{6}\right)$
14. $\frac{3}{4}\left(1\frac{3}{8}\right)$
15. $3\frac{1}{5}\left(\frac{2}{3}\right)$
16. $-\frac{5}{6}\left(2\frac{1}{2}\right)$

Multiply.

17. –2(–5.2)
18. 0.53(0.04)
19. (–7)(–3.9)
20. –2(8.13)
21. 0.02(–4.62)
22. 0.5(–7.8)
23. (–0.41)(–8.5)
24. (–8)(6.3)
25. 15(–0.05)
26. (–3.04)(–1.7)
27. 10(–0.09)
28. (–0.8)(–0.15)

29. Travis painted for $6\frac{2}{3}$ hours. He received \$27 an hour for his work. How much was Travis paid for doing this painting job?

____

Name ______________________ Date ______________ Class ______________

LESSON 2-4

Practice C

Multiplying Rational Numbers

Multiply. Write each answer in simplest form.

1. $10\left(\frac{4}{5}\right)$
2. $-12\left(\frac{7}{24}\right)$
3. $-11\left(\frac{5}{22}\right)$
4. $-18\left(\frac{5}{36}\right)$
5. $\frac{14}{28}\left(\frac{7}{42}\right)$
6. $\frac{25}{64}\left(\frac{16}{75}\right)$
7. $-\frac{14}{19}\left(\frac{38}{70}\right)$
8. $-\frac{5}{27}\left(-\frac{9}{35}\right)$
9. $\frac{9}{20}\left(\frac{36}{81}\right)$
10. $1\frac{7}{10}\left(-\frac{5}{17}\right)$
11. $\frac{39}{50}\left(-\frac{35}{117}\right)$
12. $1\frac{1}{3}\left(-\frac{63}{168}\right)$
13. $-3\left(2\frac{32}{64}\right)$
14. $\frac{2}{38}\left(9\frac{1}{2}\right)$
15. $-\frac{16}{24}\left(1\frac{24}{32}\right)$
16. $3\frac{1}{4}\left(\frac{8}{48}\right)$

Multiply.

17. 15(–13.5)
18. 6.34(1.08)
19. (–19)(–11.82)
20. 8.5(16.42)
21. 3.08(–4.38)
22. 2.8(7.15)
23. (–2.25)(–2.25)
24. (16)(2.001)

25. Mrs. Johnson harvested 107 pounds of tomatoes from her garden. She sold them for $0.85 a pound. How much did she receive from selling all the tomatoes?

____

26. A store is having a clearance sale of $\frac{1}{3}$ off the regular price. How much will be saved on a jacket with a regular price of $154.35? What will be the sale price of the jacket?

____

Name ______________________ Date ______________ Class ______________

LESSON 2-4

Review for Mastery

Multiplying Rational Numbers

To model $\frac{1}{3} \times \frac{3}{4}$.

Divide a square into 4 equal parts. Lightly shade 3 of the 4.

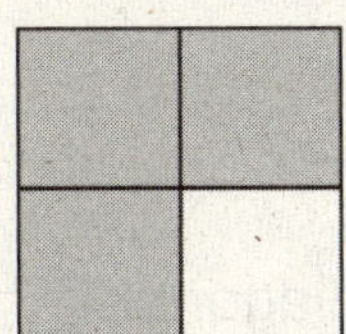

Darken 1 of the 3 shaded parts.

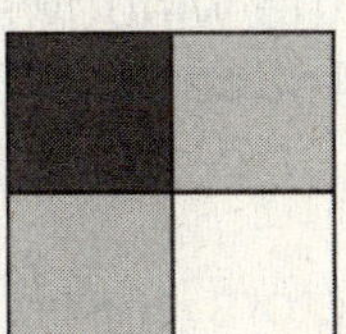

Compare the 1 darkened part to the original 4.

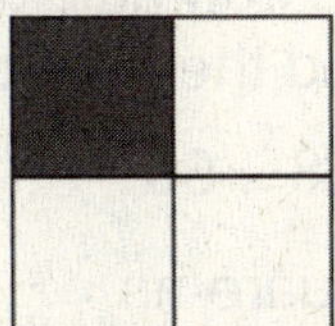

$$\frac{1}{3} \times \frac{3}{4} = \frac{1}{4}$$

Model each multiplication. Write the result.

1. 

$\frac{1}{2} \times \frac{2}{4} =$ ________

2. 

$\frac{3}{4} \times \frac{4}{6} =$ ________

3. 

$\frac{2}{3} \times \frac{3}{9} =$ ________

To multiply fractions:

- Cancel common factors, one in a numerator and the other in a denominator.
- Multiply the remaining factors in the numerator and in the denominator.
- If the signs of the factors are the same, the product is positive. If the signs of the factors are different, the product is negative.

$$\frac{\overset{1}{\cancel{3}}}{\underset{1}{\cancel{4}}} \times \frac{\overset{2}{\cancel{8}}}{\underset{3}{\cancel{9}}} = \frac{1 \times 2}{1 \times 3} = \frac{2}{3}$$

Multiply. Answer in simplest form.

4. $\frac{1}{2} \times \frac{4}{9} =$ ________

5. $\frac{2}{3} \times \frac{6}{7} =$ ________

6. $\frac{3}{5} \times \frac{15}{17} =$ ________

7. $\frac{2}{3} \times \left(-\frac{9}{10}\right) =$ ________

8. $\left(-\frac{2}{9}\right) \times \frac{27}{40} =$ ________

9. $\left(-\frac{4}{7}\right) \times \left(-\frac{21}{8}\right) =$ ________

Name ______________________________ Date __________________ Class __________________

LESSON 2-4

Challenge

Curtains

Variations of modern long multiplication were introduced into Europe by a 13th-century Italian, Leonardo of Pisa (Fibonacci). Many multiplication techniques can be traced to a book called *Lilaviti*, written by Bhaskara for his daughter in 12th-century India.

Here's how to do multiplication by the **Gelosia Method,** named after *jalousie*, the iron grill Italians placed over the windows. The method is also called the **Lattice Method of Multiplication.**

Consider: 38 × 56

Divide a square as shown.

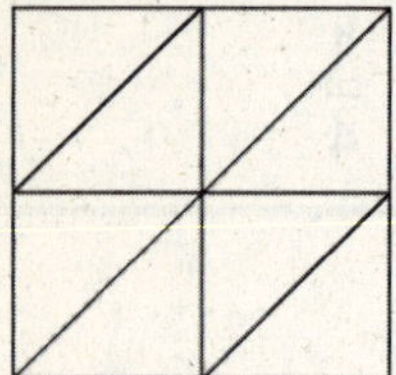

Align the factors. Insert the individual products.

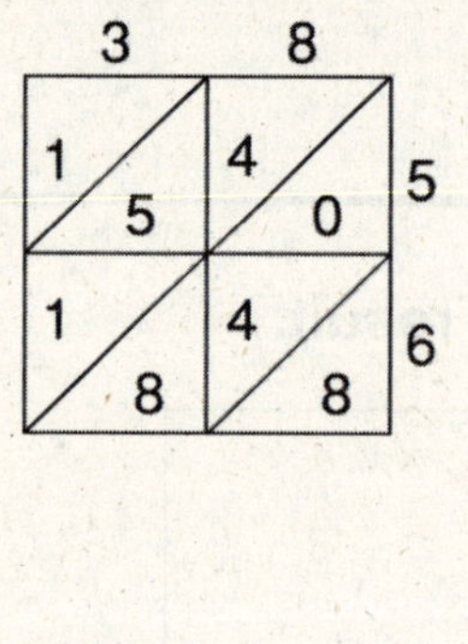

Sum each diagonal; begin with lower right. As needed, carry numbers into next diagonal sum.

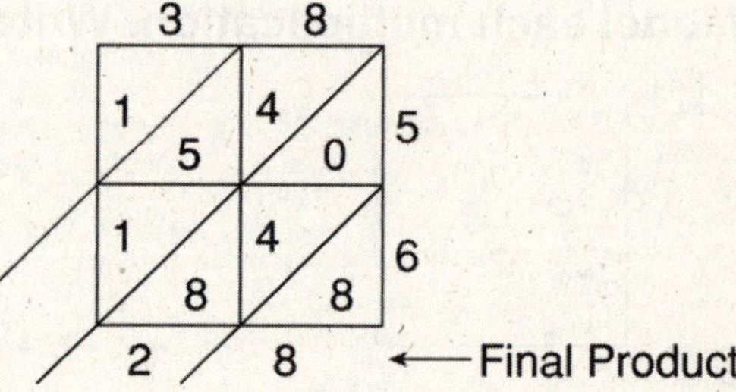

So, 38 × 56 = 2128.

Use the Gelosia Method to multiply.
Verify results by your usual method.

1.

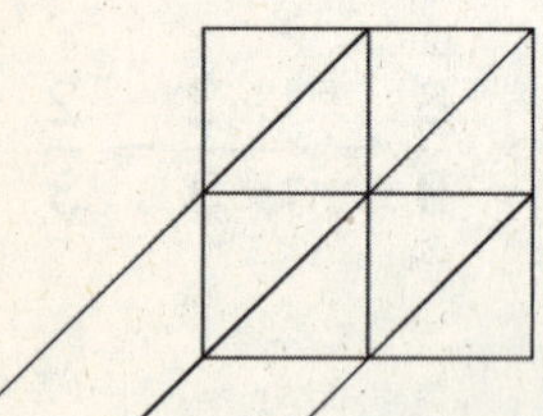

39 × 47 = ________

2.

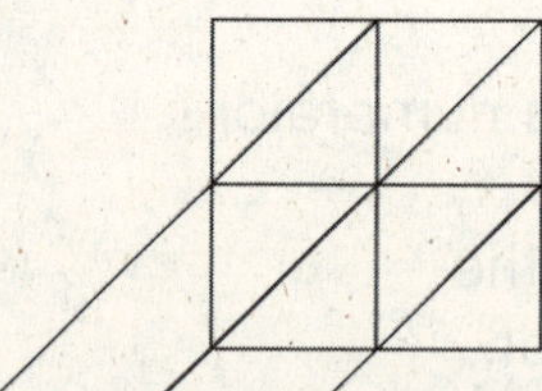

68 × 73 = ________

3.

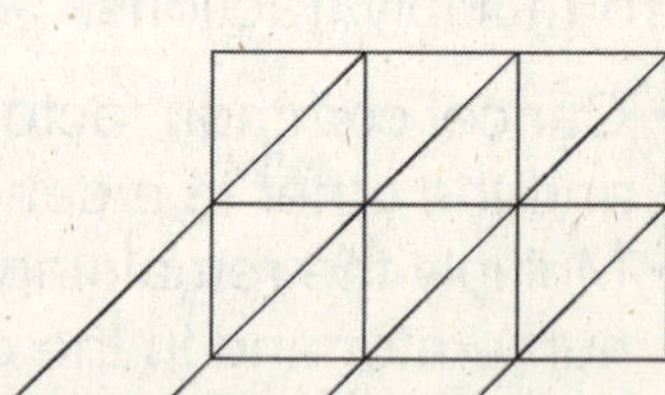

358 × 64 = ________

Name ______________________ Date ____________ Class ____________

LESSON 2-4

Problem Solving

Multiplying Rational Numbers

Use the table at the right.

Average World Births and Deaths per Second in 2001

Births	$4\frac{1}{5}$
Deaths	1.7

1. What was the average number of births per minute in 2001?

____

2. What was the average number of deaths per hour in 2001?

____

3. What was the average number of births per day in 2001?

____

4. What was the average number of births in $\frac{1}{2}$ of a second in 2001?

____

5. What was the average number of births in $\frac{1}{4}$ of a second in 2001?

____

Use the table below. During exercise, the target heart rate is 0.5–0.75 of the maximum heart rate. Choose the letter for the best answer.

Age	Maximum Heart Rate
13	207
14	206
15	205
20	200
25	195

Source: American Heart Association

6. What is the target heart rate range for a 14 year old?
   A 7–10.5
   B 103–154.5
   C 145–166
   D 206–255

7. What is the target heart rate range for a 20 year old?
   F 100–150
   G 125–175
   H 150–200
   J 200–250

8. What is the target heart rate range for a 25 year old?
   A 25–75
   B 85–125
   C 97.5–146.25
   D 195–250

Name ______________________ Date ______________ Class ______________

LESSON 2-4

Reading Strategies

Use a Visual Model

This rectangle will help you understand how to find the product of $\frac{1}{2} \cdot \frac{1}{3}$. First, $\frac{1}{2}$ of the rectangle was shaded. Then, the rectangle was divided horizontally into thirds. Then, $\frac{1}{3}$ was shaded. The overlap of the shading shows the product of $\frac{1}{2} \cdot \frac{1}{3}$.

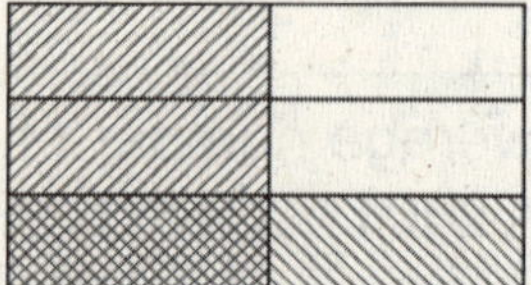

1. Into how many parts is the rectangle divided? What fractional part of the rectangle is each of these parts? ______________
2. What fractional part of the rectangle has shading that overlaps? ______________
3. Multiply the numerators and the denominators of the given fractions. = ______________
4. Use the rectangle to draw a model for the problem $\frac{1}{4} \cdot \frac{1}{2}$.

5. Draw lines from top to bottom to divide the rectangle into fourths. Shade one-fourth of the rectangle.
6. Draw a line across the rectangle to divide it into halves. Into how many parts is the rectangle now divided? ______________
7. Shade one of the halves.
8. What fractional part of the rectangle was shaded twice? ______________
9. Multiply the numerators and denominators. ______________

Name ______________________ Date ______________ Class ______________

LESSON 2-4

Puzzles, Twisters & Teasers

Egg-zactly Correct!

Multiply. Write each answer in its simplest form. Then solve the riddle using the answers.

Y $\frac{7}{8}\left(\frac{3}{5}\right) =$ ______________

N $3\left(2\frac{1}{5}\right) =$ ______________

T $-2\left(\frac{9}{16}\right) =$ ______________

E $6\left(\frac{2}{3}\right) =$ ______________

I $-5\left(1\frac{3}{4}\right) =$ ______________

W $2\left(\frac{7}{8}\right) =$ ______________

C $\frac{3}{4}\left(-\frac{1}{8}\right) =$ ______________

O $\frac{6}{8}\left(\frac{2}{5}\right) =$ ______________

K $1\frac{2}{3}\left(\frac{5}{6}\right) =$ ______________

L $-\frac{1}{3}\left(-\frac{4}{7}\right) =$ ______________

What do you call a city with a million eggs?

___ ___ ___   ___ ___ ___ ___

$6\frac{3}{5}$ 4 $1\frac{3}{4}$ $\frac{21}{40}$ $\frac{3}{10}$ $\frac{4}{21}$ $1\frac{7}{18}$

___ ___ ___ ___

$-\frac{3}{32}$ $-8\frac{3}{4}$ $-1\frac{1}{8}$ $\frac{21}{40}$

Name ______________________ Date ______________ Class ______________

LESSON 2-5

Practice A

Dividing Rational Numbers

Divide. Write each answer in simplest form.

1. $\frac{1}{4} \div \frac{3}{8}$ ____________

2. $-\frac{2}{3} \div \frac{5}{9}$ ____________

3. $\frac{1}{6} \div \frac{1}{3}$ ____________

4. $\frac{3}{4} \div \left(-\frac{1}{8}\right)$ ____________

5. $\frac{1}{9} \div \frac{1}{3}$ ____________

6. $\frac{2}{5} \div \frac{4}{7}$ ____________

7. $-\frac{3}{5} \div \frac{6}{7}$ ____________

8. $-\frac{3}{8} \div \left(-\frac{5}{6}\right)$ ____________

9. $1\frac{2}{5} \div 1\frac{1}{2}$ ____________

10. $-\frac{3}{4} \div 9$ ____________

11. $-2\frac{1}{3} \div \frac{1}{4}$ ____________

12. $-\frac{5}{8} \div 5$ ____________

Find each quotient.

13. 1.53 ÷ 0.3 ____________

14. 5.14 ÷ 0.2 ____________

15. 10.05 ÷ 0.05 ____________

16. 5.28 ÷ 0.4 ____________

17. 6.54 ÷ 0.03 ____________

18. 29.45 ÷ 0.005 ____________

19. 8.58 ÷ 0.06 ____________

20. 1.61 ÷ 0.7 ____________

Evaluate each expression for the given value of the variable.

21. $\frac{10}{x}$ for $x = 0.05$ ____________

22. $\frac{9.12}{x}$ for $x = -0.2$ ____________

23. $\frac{42.42}{x}$ for $x = 1.4$ ____________

24. Mr. Chen has a 76-in. space to stack books. Each book is $6\frac{1}{3}$ in. tall. How many books can he stack in the space?

____

Name ______________________ Date ______________ Class ______________

LESSON 2-5

Practice B

Dividing Rational Numbers

Divide. Write each answer in simplest form.

1. $\frac{1}{5} \div \frac{3}{10}$ ________

2. $-\frac{5}{8} \div \frac{3}{4}$ ________

3. $\frac{1}{4} \div \frac{1}{8}$ ________

4. $-\frac{2}{3} \div \frac{4}{15}$ ________

5. $1\frac{2}{9} \div 1\frac{2}{3}$ ________

6. $-\frac{7}{10} \div \left(\frac{2}{5}\right)$ ________

7. $\frac{6}{11} \div \frac{3}{22}$ ________

8. $\frac{4}{9} \div \left(-\frac{8}{15}\right)$ ________

9. $\frac{3}{8} \div -15$ ________

10. $-\frac{5}{6} \div 12$ ________

11. $6\frac{1}{2} \div 1\frac{5}{8}$ ________

12. $-\frac{9}{10} \div 6$ ________

Find each quotient.

13. 24.35 ÷ 0.5 ________

14. 2.16 ÷ 0.04 ________

15. 3.16 ÷ 0.02 ________

16. 7.32 ÷ 0.3 ________

17. 87.36 ÷ 0.6 ________

18. 79.36 ÷ 0.8 ________

19. 4.27 ÷ 0.007 ________

20. 63.81 ÷ 0.9 ________

21. 1.23 ÷ 0.003 ________

22. 62.46 ÷ 0.09 ________

23. 21.12 ÷ 0.4 ________

24. 82.68 ÷ 0.06 ________

Evaluate each expression for the given value of the variable.

25. $\frac{18}{x}$ for $x = 0.12$ ________

26. $\frac{10.8}{x}$ for $x = 0.03$ ________

27. $\frac{9.18}{x}$ for $x = -1.2$ ________

28. A can of fruit contains $3\frac{1}{2}$ cups of fruit. The suggested serving size is $\frac{1}{2}$ cup. How many servings are in the can of fruit?

____

Name ______________________ Date ______________ Class ______________

LESSON 2-5

Practice C

Dividing Rational Numbers

Divide. Write each answer in simplest form.

1. $\frac{10}{15} \div \frac{8}{25}$

2. $1\frac{3}{18} \div 2\frac{1}{3}$

3. $\frac{6}{13} \div \frac{18}{26}$

4. $-\frac{12}{21} \div \frac{3}{14}$

5. $-\frac{3}{20} \div \left(\frac{15}{35}\right)$

6. $-\frac{27}{32} \div 3$

7. $\frac{36}{41} \div \frac{9}{82}$

8. $2\frac{12}{34} \div 1\frac{3}{17}$

Find each quotient.

9. $3.15 \div 0.05$

10. $26.008 \div 0.4$

11. $-983.1 \div (-0.3)$

12. $1.44 \div 0.16$

13. $236.4 \div 0.0012$

14. $-10.08 \div 0.005$

15. $2.253 \div 0.15$

16. $-1.161 \div 0.18$

Evaluate each expression for the given value of the variable.

17. $\frac{2.25}{x}$ for $x = -0.009$

18. $\frac{-234.72}{x}$ for $x = 3.6$

19. $\frac{-4330.8}{x}$ for $x = -8.02$

20. Emma bought $2\frac{1}{2}$ yards of cording for the trim around the edge of a square pillow. How much will she use for each side of the pillow?

21. Sean has a loan of $8804.46 including interest. He makes payments of $209.63 each month on the simple interest loan. How many months will it take for Sean to repay his loan?

Name ________________ Date ________________ Class ________________

LESSON 2-5

Review for Mastery

Dividing Rational Numbers

To write the **reciprocal** of a fraction, interchange the numerator and denominator.

$\frac{2}{3}$ ⇄ $\frac{3}{2}$

Fraction Reciprocal

The product of a number and its reciprocal is 1.

$\frac{2}{3} \times \frac{3}{2} = 1$

Write the reciprocal of each rational number.

1. The reciprocal of $\frac{3}{5}$ is: ________________

2. The reciprocal of 6 is: ________________

3. The reciprocal of $2\frac{1}{3}$ is: ________________

To divide by a fraction, multiply by its reciprocal.

$\frac{2}{3} \div 6$

$\frac{2}{3} \times \frac{1}{6}$

$\frac{{}^{1}\cancel{2} \times 1}{3 \times \cancel{6}_{3}} = \frac{1}{9}$

$\frac{3}{5} \div \frac{9}{10}$

$\frac{3}{5} \times \frac{10}{9}$

$\frac{\overset{1}{\cancel{3}} \times \overset{2}{\cancel{10}}}{\underset{1}{\cancel{5}} \times \underset{3}{\cancel{9}}} = \frac{2}{3}$

Complete to divide and simplify.

4. $\frac{3}{8} \div 12 = \frac{3}{8} \times$ ______ = ________

5. $\frac{4}{3} \div 16 = \frac{4}{3} \times$ ______ = ________

6. $\frac{5}{7} \div \frac{20}{21} = \frac{5}{7} \times$ ______ = ________

7. $-\frac{3}{4} \div \left(\frac{9}{8}\right) = -\frac{3}{4} \times$ ______ = ________

Change a decimal divisor to a whole number. Using the number of places in the divisor, move the decimal point to the right in both the divisor and the dividend.

$0.7\overline{)4.34} \rightarrow 0.7.\overline{)4.3.4} \rightarrow 7\overline{)43.4}$ (quotient 6.2)

Rewrite each division with a whole-number divisor. Then, do the division.

8. $0.6\overline{)1.14} \rightarrow$ ________ = ______

9. $0.3\overline{)4.56} \rightarrow$ ________ = ______

10. $0.02\overline{)7.12} \rightarrow$ ________ = ______

11. $0.08\overline{)57.28} \rightarrow$ ________ = ______

Name ______________________ Date ______________ Class ______________

LESSON 2-5

Challenge

A New License to Operate

You can invent new operations based on the familiar operations of addition, subtraction, multiplication, and division, and the familiar order of operations.

If $a \,\Delta\, b = \frac{a+b}{2}$ where a and b represent any rational numbers,

then $3 \,\Delta\, 5 = \frac{3+5}{2} = 4$.

Use the given definition of operation Δ to evaluate each expression.

1. $\frac{1}{2} \,\Delta\, (-10) =$ ____________

2. $\frac{100 \,\Delta\, (-10)}{10} =$ ____________

3. $4 \,\Delta\, 6 \,\Delta\, 3 =$ ____________

4. $[5.5 \,\Delta\, (-6)] + [-6 \,\Delta\, 5.5] =$ ____________

Use the operation shown to answer each question. $\boxed{\frac{a}{c} \,\Big|\, \frac{b}{d}} = ac - bd$

5. $\boxed{\frac{1}{3} \,\Big|\, \frac{8}{4}} =$ ____________

6. $\boxed{\frac{-2}{3} \,\Big|\, \frac{3}{-2}} =$ ____________

7. If $\boxed{\frac{1}{x} \,\Big|\, \frac{3}{2}} = 18$, then $x =$ ____________

8. If $\boxed{\frac{6}{x} \,\Big|\, \frac{2}{x}} = 12$, then $x =$ ____________

Use the operation shown to answer each question. $a \nwarrow\!\searrow b = \frac{a^2}{b^2}$

9. $(3 \nwarrow\!\searrow 5) =$ ____________

10. $(1 \nwarrow\!\searrow 8) - (5 \nwarrow\!\searrow 8) =$ ____________

11. $(1 \nwarrow\!\searrow 3) \times (3 \nwarrow\!\searrow 6) =$ ____________

12. $(1 \nwarrow\!\searrow 10)^2 =$ ____________

If ↲*n*↲ means 1 less than the number of digits in the integer *n*, then, for example, ↲77↲ = 1 since 77 has 2 digits.

Use the definition of ↲*n*↲ to answer each question.

13. If *n* is a positive integer less than 100, what is the greatest value for ↲*n*↲? ____________

14. If *n* is a positive integer less than 1001, what is the greatest value for ↲*n*↲? ____________

15. If *n* has 100 digits, what is the value of ↲↲*n*↲↲? Explain.

____

____

Name ______________________ Date ______________ Class ______________

LESSON 2-5

Problem Solving

Dividing Rational Numbers

Use the table at the right that shows the maximum speed over a quarter mile of different animals. Find the time it takes each animal to travel one-quarter mile at top speed. Round to the nearest thousandth.

Maximum Speeds of Animals

Animal	**Speed** (mph)
Quarter horse	47.50
Greyhound	39.35
Human	27.89
Giant tortoise	0.17
Three-toed sloth	0.15

1. Quarter horse

____

2. Greyhound

____

3. Human

____

4. Giant tortoise

____

5. Three-toed sloth

____

Choose the letter for the best answer.

6. A piece of ribbon is $1\frac{7}{8}$ inches long. If the ribbon is going to be divided into 15 pieces, how long should each piece be?

   A $\frac{1}{8}$ in.

   B $\frac{1}{15}$ in.

   C $\frac{2}{3}$ in.

   D $28\frac{1}{8}$ in.

7. The recorded rainfall for each day of a week was 0 in., $\frac{1}{4}$ in., $\frac{3}{4}$ in., 1 in., 0 in., $1\frac{1}{4}$ in., $1\frac{1}{4}$ in. What was the average rainfall per day?

   F $\frac{9}{10}$ in.

   G $\frac{9}{14}$ in.

   H $\frac{7}{8}$ in.

   J $4\frac{1}{2}$ in.

8. A drill bit that is $\frac{7}{32}$ in. means that the hole the bit makes has a diameter of $\frac{7}{32}$ in. Since the radius is half of the diameter, what is the radius of a hole drilled by a $\frac{7}{32}$ in. bit?

   A $\frac{14}{32}$ in.

   B $\frac{7}{32}$ in.

   C $\frac{9}{16}$ in.

   D $\frac{7}{64}$ in.

9. A serving of a certain kind of cereal is $\frac{2}{3}$ cup. There are 12 cups of cereal in the box. How many servings of cereal are in the box?

   F 18

   G 15

   H 8

   J 6

Name ______________________ Date ______________ Class ______________

LESSON 2-5

Reading Strategies

Focus on Vocabulary

The word **reciprocal** means an exchange. When two friends exchange gifts, you might think of the gifts as "switching places." In the reciprocal of a fraction, the numerator and denominator exchange places.

Fraction	**Reciprocal**
$\frac{2}{3}$	$\frac{3}{2}$
$\frac{4}{5}$	$\frac{5}{4}$
$\frac{8}{1}$	$\frac{1}{8}$

1. What does the word *reciprocal* mean? ______________

2. What is the reciprocal of $\frac{7}{8}$? ______________

3. What is the reciprocal of $\frac{6}{5}$? ______________

The product of a fraction and its reciprocal is always 1.

Fraction • Reciprocal = Product

$\frac{2}{3} \cdot \frac{3}{2} = \frac{6}{6} = 1$

$\frac{4}{5} \cdot \frac{5}{4} = \frac{20}{20} = 1$

$\frac{1}{8} \cdot \frac{8}{1} = \frac{8}{8} = 1$

4. What is the product of $\frac{1}{7} \cdot \frac{7}{1}$? ______________

5. What is the product of $\frac{2}{6}$ and its reciprocal? ______________

6. What is the reciprocal of $\frac{1}{2}$? ______________

7. What is the product of $\frac{1}{2} \times 2$? ______________

Name ______________________ Date ______________ Class ______________

LESSON 2-5

Puzzles, Twisters & Teasers

Hungry for Knowledge!

Solve each equation. Round answers to the nearest tenth. Then use the answers to solve the riddle.

R $17.78 \div 0.7 =$ ______________

L $\frac{1}{2} \div \frac{1}{4} =$ ______________

E $10.08 \div 0.6 =$ ______________

U $24 \div \frac{3}{4} =$ ______________

N $3.72 \div 0.3 =$ ______________

C $10\frac{1}{2} \div 3\frac{1}{2} =$ ______________

I $14.08 \div 0.8 =$ ______________

H $3\frac{1}{3} \div 1\frac{1}{5} =$ ______________

D $9.36 \div 0.03 =$ ______________

A $6.52 \div 0.004 =$ ______________

What are two things you can't eat for breakfast?

___	___	___	___	___		___	___	___
2	32	12.4	3	2.8		1630	12.4	312

___	___	___	___	___	___
312	17.6	12.4	12.4	16.8	25.4

Name ______________________ Date ______________ Class ______________

LESSON 2-6

Practice A

Adding and Subtracting with Unlike Denominators

Name a common denominator for each sum or difference. Do not solve.

1. $\frac{1}{2}+\frac{3}{4}$ ________

2. $\frac{1}{3}+\frac{4}{9}$ ________

3. $\frac{2}{3}-\frac{3}{8}$ ________

4. $\frac{1}{2}-\frac{1}{6}$ ________

Add or subtract. Write each answer in simplest form.

5. $\frac{1}{5}+\frac{1}{2}$ ________

6. $\frac{3}{4}+\frac{5}{6}$ ________

7. $\frac{7}{10}-\frac{1}{5}$ ________

8. $\frac{3}{14}-\frac{4}{7}$ ________

9. $3\frac{1}{3}-1\frac{1}{6}$ ________

10. $\frac{1}{4}+\frac{1}{2}$ ________

11. $4\frac{1}{3}-2\frac{7}{9}$ ________

12. $2\frac{3}{5}+\left(-1\frac{7}{10}\right)$ ________

13. $\frac{2}{5}-\frac{1}{2}$ ________

14. $2\frac{1}{3}+1\frac{1}{2}$ ________

15. $3\frac{1}{4}+\left(-1\frac{5}{6}\right)$ ________

16. $\frac{3}{4}-\frac{11}{12}$ ________

Evaluate each expression for the given value of the variable.

17. $1\frac{7}{8}+x$ for $x=-2\frac{3}{4}$ ________

18. $x-\frac{2}{3}$ for $x=\frac{1}{6}$ ________

19. $x-\frac{1}{2}$ for $x=\frac{7}{8}$ ________

20. $2\frac{2}{3}+x$ for $x=-1\frac{5}{9}$ ________

21. $x-\frac{2}{5}$ for $x=\frac{9}{10}$ ________

22. $x-\frac{6}{7}$ for $x=\frac{1}{2}$ ________

23. Mr. Martanarie bought a new lamp and lamppost for his home. The pole was $6\frac{5}{8}$ ft tall and the lamp was $1\frac{1}{4}$ ft in height. How tall were the lamp and post together?

____

Name ______________________ Date ______________ Class ______________

LESSON 2-6

Practice B

Adding and Subtracting with Unlike Denominators

Add or subtract.

1. $\frac{2}{3}+\frac{1}{2}$ ________

2. $\frac{3}{5}+\frac{1}{3}$ ________

3. $\frac{3}{4}-\frac{1}{3}$ ________

4. $\frac{1}{2}-\frac{5}{9}$ ________

5. $\frac{5}{16}-\frac{5}{8}$ ________

6. $\frac{7}{9}+\frac{5}{6}$ ________

7. $\frac{7}{8}-\frac{1}{4}$ ________

8. $\frac{5}{6}-\frac{3}{8}$ ________

9. $2\frac{7}{8}+3\frac{5}{12}$ ________

10. $1\frac{2}{9}+2\frac{1}{18}$ ________

11. $3\frac{2}{3}-1\frac{3}{5}$ ________

12. $1\frac{5}{6}+\left(-2\frac{3}{4}\right)$ ________

13. $8\frac{1}{3}-3\frac{5}{9}$ ________

14. $5\frac{1}{3}+1\frac{11}{12}$ ________

15. $7\frac{1}{4}+\left(-2\frac{5}{12}\right)$ ________

16. $5\frac{2}{5}-7\frac{3}{10}$ ________

Evaluate each expression for the given value of the variable.

17. $2\frac{3}{8}+x$ for $x=1\frac{5}{6}$ ________

18. $x-\frac{2}{5}$ for $x=\frac{1}{3}$ ________

19. $x-\frac{3}{10}$ for $x=\frac{3}{7}$ ________

20. $1\frac{5}{8}+x$ for $x=-2\frac{1}{6}$ ________

21. $x-\frac{3}{4}$ for $x=\frac{1}{6}$ ________

22. $x-\frac{3}{10}$ for $x=\frac{1}{2}$ ________

23. Ana worked $6\frac{1}{2}$ h on Monday, $5\frac{3}{4}$ h on Tuesday and $7\frac{1}{6}$ h on Friday. How many total hours did she work these three days?

____

Name ______________________ Date ______________ Class ______________

LESSON 2-6

Practice C

Adding and Subtracting with Unlike Denominators

Add or subtract.

1. $\frac{7}{12}+\frac{5}{9}$

2. $\frac{7}{10}+\frac{4}{15}$

3. $\frac{7}{8}-\frac{11}{12}$

4. $\frac{15}{16}-\frac{5}{32}$

5. $1\frac{1}{18}+\left(-4\frac{15}{24}\right)$

6. $8\frac{3}{20}-3\frac{7}{30}$

7. $8\frac{7}{10}+\left(-6\frac{2}{25}\right)$

8. $5\frac{11}{15}-7\frac{5}{6}$

Evaluate each expression for the given value of the variable.

9. $10\frac{6}{25}+x$ for $x=-2\frac{3}{5}$

10. $x-\frac{5}{18}$ for $x=\frac{5}{6}$

11. $1\frac{5}{21}+x$ for $x=-5\frac{6}{7}$

12. $x-\frac{8}{11}$ for $x=\frac{13}{22}$

13. $14\frac{1}{15}+x$ for $x=-9\frac{3}{10}$

14. $x-\frac{7}{9}$ for $x=\frac{2}{27}$

15. A carpenter cuts a piece of wood that is $9\frac{5}{16}$ ft long into two pieces. One piece is $5\frac{3}{8}$ ft long. How long is the other piece?

16. Before April 9, 2001, when the U.S. Securities and Exchange Commission ordered all U.S. stock markets to report stocks in decimals, the price of stock was reported in fractions. Under the fractional reporting system, what was the change in stock price if a stock opened the day at $31\frac{1}{8}$ and closed the day at $28\frac{5}{32}$?

Name ______________________ Date ______________ Class ______________

LESSON 2-6

Review for Mastery

Adding and Subtracting with Unlike Denominators

To model $\frac{1}{2} + \frac{1}{3}$, use two rectangles of the same size and shape.

A. 1st rectangle: Shade $\frac{1}{2}$ vertically.

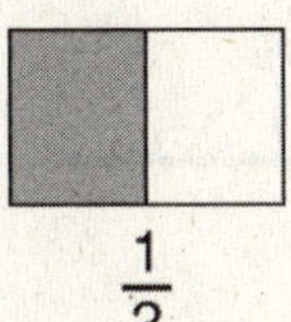

$\frac{1}{2}$

B. 2nd rectangle: Shade $\frac{1}{3}$ horizontally.

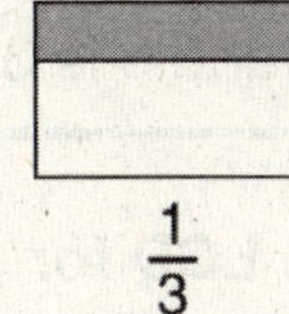

$\frac{1}{3}$

C. Separate the shaded portions into parts of equal size.

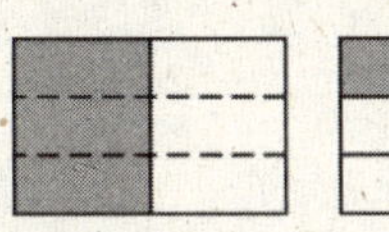

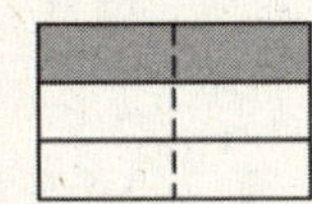

$\frac{1}{2} = \frac{3}{6}$ $\frac{1}{3} = \frac{2}{6}$

D. Use a new rectangle to show the sum.

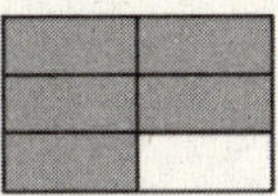

$\frac{1}{2} + \frac{1}{3} = \frac{3}{6} + \frac{2}{6} = \frac{5}{6}$

Model $\frac{1}{2} + \frac{2}{5}$. Write the result.

1.

Model $\frac{1}{3} + \frac{3}{5}$. Write the result.

2.

Name ______________________ Date ______________ Class ______________

LESSON 2-6

Review for Mastery

Adding and Subtracting with Unlike Denominators (continued)

To add fractions with different denominators, first write the fractions with common denominators. To find the LCD of denominators 5 and 6, list the multiples of each.

Multiples of 5: 5, 10, 15, 20, 25, (30)
Multiples of 6: 12, 18, 24, (30)
So, the LCD of 5 and 6 is 30.

Complete to find the LCD for each set of denominators.

3. The LCD of 6 and 4 is: ______________

   Multiples of 6: ______________

   Multiples of 4: ______________

4. The LCD of 3 and 7 is: ______________

   Multiples of 3: ______________

   Multiples of 7: ______________

To add fractions with different denominators:

Add: $\frac{1}{2} + \frac{1}{3} = \frac{1 \cdot 3}{2 \cdot 3} = \frac{3}{6}$

$\frac{1 \cdot 2}{3 \cdot 2} = \frac{2}{6}$

$= \frac{5}{6}$

Complete to add fractions. Simplify.

5. $\frac{1}{4} = \frac{\quad}{20}$

   $+\frac{3}{5} = \frac{\quad}{20}$

   $=$ ______

6. $\frac{3}{4} = \frac{\quad}{16}$

   $+\frac{5}{16} = \frac{\quad}{16}$

   $=$ ______ $=$ ______

7. $5\frac{1}{3} = 5\frac{\quad}{24}$

   $+2\frac{5}{8} = 2\frac{\quad}{24}$

   $=$ ______

Add or subtract fractions. Simplify.

8. $\frac{1}{4} + \frac{7}{20} =$

9. $\frac{4}{9} - \frac{1}{5} =$

10. $\frac{8}{15} - \frac{1}{4} =$

Name ______________________ Date ______________ Class ______________

LESSON 2-6

Challenge

Please Repeat That.

A decimal that repeats one digit is equivalent to a fraction with denominator 9.

$0.\overline{1} = \frac{1}{9}$ $0.\overline{2} = \frac{2}{9}$ $0.\overline{5} = \frac{5}{9}$

A decimal that repeats two digits is equivalent to a fraction with denominator 99.

$0.\overline{43} = \frac{43}{99}$ $0.\overline{61} = \frac{61}{99}$ $0.\overline{38} = \frac{38}{99}$

The pattern continues so that $0.\overline{681} = \frac{681}{999}$ and $0.\overline{24793} = \frac{24{,}793}{99{,}999}$.

Use a calculator to write each decimal equivalent.

1. $\frac{1}{90} =$ ______________
2. $\frac{21}{990} =$ ______________
3. $\frac{358}{9990} =$ ______________

Predict the decimal equivalent of each fraction. Verify your results on a calculator.

4. $\frac{4}{90} =$ ______________
5. $\frac{62}{990} =$ ______________
6. $\frac{617}{9990} =$ ______________

Write each fractional equivalent and simplify.

7. $0.\overline{7} =$ ______________
8. $0.0\overline{8} =$ ______________
9. $0.00\overline{24} =$ ______________

When one digit repeats but does not begin in the first decimal place, and the digit in the first place is other than 0, you must add fractions.

$$0.3\overline{7} = 0.3 + 0.0\overline{7}$$
$$= \frac{3}{10} + \frac{7}{90}$$
$$= \frac{27}{90} + \frac{7}{90} = \frac{34}{90} = \frac{17}{45}$$

Write each repeating decimal as the sum of two fractions. Find the sum and simplify. Verify.

10. $0.2\overline{8} =$ ______________
11. $0.2\overline{532} =$ ______________
12. $0.1\overline{27} =$ ______________
13. $0.75\overline{483} =$ ______________

Name ______________________ Date ______________ Class ______________

LESSON 2-6

Problem Solving

Adding and Subtracting with Unlike Denominators

Write the correct answer.

1. Nick Hysong of the United States won the Olympic gold medal in the pole vault in 2000 with a jump of 19 ft $4\frac{1}{4}$ inches, or $232\frac{1}{4}$ inches. In 1900, Irving Baxter of the United States won the pole vault with a jump of 10 ft $9\frac{7}{8}$ inches, or $129\frac{7}{8}$ inches. How much higher did Hysong vault than Baxter?

   ____

2. In the 2000 Summer Olympics, Ivan Pedroso of Cuba won the Long jump with a jump of 28 ft $\frac{3}{4}$ inches, or $336\frac{3}{4}$ inches. Alvin Kraenzlein of the United States won the long jump in 1900 with a jump of 23 ft $6\frac{7}{8}$ inches, or $282\frac{7}{8}$ inches. How much farther did Pedroso jump than Kraenzlein?

   ____

3. A recipe calls for $\frac{1}{8}$ cup of sugar and $\frac{3}{4}$ cup of brown sugar. How much total sugar is added to the recipe?

   ____

4. The average snowfall in Norfolk, VA for January is $2\frac{3}{5}$ inches, February $2\frac{9}{10}$ inches, March 1 inch, and December $\frac{9}{10}$ inches. If these are the only months it typically snows, what is the average snowfall per year?

   ____

Use the table at the right that shows the average snowfall per month in Vail, Colorado.

5. What is the average annual snowfall in Vail, Colorado?

   A $15\frac{13}{20}$ in.

   B 153 in.

   C $187\frac{1}{10}$ in.

   D $187\frac{4}{5}$ in.

6. The peak of the skiing season is from December through March. What is the average snowfall for this period?

   F $30\frac{19}{20}$ in.

   G $123\frac{3}{5}$ in.

   H $123\frac{4}{5}$ in.

   J 127 in.

Average Snowfall in Vail, CO

Month	Snowfall (in.)	Month	Snowfall (in.)
Jan	$36\frac{7}{10}$	July	0
Feb	$35\frac{7}{10}$	August	0
March	$25\frac{2}{5}$	Sept	1
April	$21\frac{1}{5}$	Oct	$7\frac{4}{5}$
May	4	Nov	$29\frac{7}{10}$
June	$\frac{3}{10}$	Dec	26

Name ______________________________ Date ______________ Class ______________

LESSON 2-6

Reading Strategies

Use a Graphic Aid

It is easy to add and subtract fractions with **common denominators.**

3 eighths + 4 eighths = 7 eighths 8 ninths – 3 ninths = 5 ninths

$\frac{3}{8} + \frac{4}{8} = \frac{7}{8}$ $\frac{8}{9} - \frac{3}{9} = \frac{5}{9}$

Adding fractions with unlike denominators requires more steps. The picture below will help you understand adding fractions with unlike denominators. $\frac{1}{2} + \frac{1}{4} = ?$

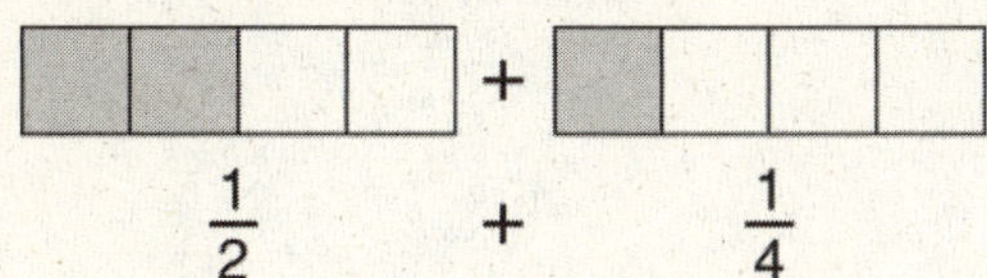

$\frac{1}{2}$ + $\frac{1}{4}$

In order to add $\frac{1}{2} + \frac{1}{4}$, you must find a common denominator.

1. What are the denominators in this problem? ______________
2. To find a common denominator, one-half can be changed into fourths. How many fourths are there in one-half? ______________
3. Change $\frac{1}{2}$ to fourths. ______________
4. You can now add, because you have a common denominator. ______________

To subtract fractions with unlike denominators, you must find a common denominator. The picture below will help you understand finding a common denominator. $\frac{5}{6} - \frac{1}{3} = ?$

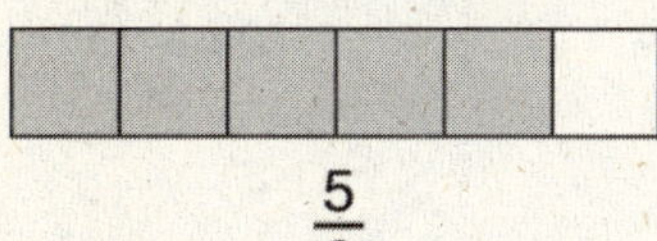

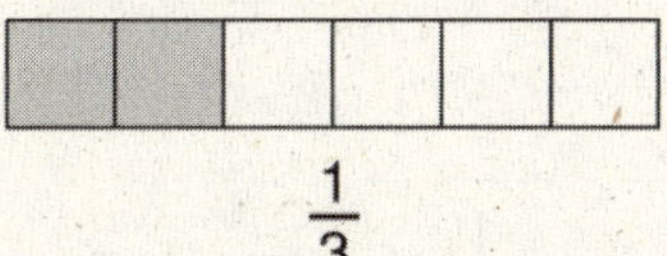

$\frac{5}{6}$ $\frac{1}{3}$

5. What are the denominators in this problem? ______________

To find a common denominator, you will change to sixths.

6. How many sixths are in one-third? Write the fraction. ______________
7. You can now subtract the fractions. ______________

Name ______________________ Date ______________ Class ______________

LESSON 2-6

Puzzles, Twisters & Teasers

Just a Tad Difficult!

Add or subtract to solve each equation. Then use the answers to solve the riddle.

M $\frac{5}{12}+\frac{3}{7}=$ ______________

T $\frac{1}{5}+\frac{7}{9}=$ ______________

O $\frac{15}{16}-\frac{9}{10}=$ ______________

H $\frac{1}{3}+1\frac{1}{12}=$ ______________

R $2\frac{1}{5}+1\frac{8}{9}=$ ______________

E $\frac{5}{8}+\frac{1}{6}=$ ______________

K $\frac{5}{16}+\frac{2}{7}=$ ______________

C $2\frac{1}{3}-4\frac{7}{9}=$ ______________

A $\frac{3}{4}-\frac{5}{16}=$ ______________

N $1\frac{3}{4}+\left(-3\frac{13}{15}\right)=$ ______________

Where do tadpoles change into frogs?

I ___ ___ ___ ___

$-2\frac{7}{60}$ $\qquad \frac{44}{45}$ $\quad 1\frac{5}{12}$ $\quad \frac{19}{24}$

___ ___ ___ ___ ___ ___ ___ ___ ___

$-2\frac{4}{9}$ $\quad 4\frac{4}{45}$ $\quad \frac{3}{80}$ $\quad \frac{7}{16}$ $\quad \frac{67}{112}$ $\qquad 4\frac{4}{45}$ $\quad \frac{3}{80}$ $\quad \frac{3}{80}$ $\quad \frac{71}{84}$

Family Letter

2B Equations with Rational Numbers

Dear Family,

The student will learn to solve equations that involve rational numbers. As with equations involving integers, the variable must be isolated to find its value. For example:

$$\frac{1}{3}x = 7$$

$3 \times \frac{1}{3}x = 7 \times 3$ — Multiply both sides by 3.

$x = 7 \times 3$ — Simplify.

$x = 21$

Some equations involve more than one step. An example of this is:

$$\frac{k}{2} + 2 = 8$$

We can begin to understand how to solve this equation by making a model. We can substitute objects for the numbers.

$$\frac{k}{\square\square} + \square\square = \begin{matrix}\square\square\square\square\\\square\square\square\square\end{matrix}$$

$$\underline{-\square\square} \qquad \underline{-\square\square}$$

$$\frac{k}{\square\square} = \begin{matrix}\square\square\square\\\square\square\square\end{matrix}$$

$$\square\square \times \frac{k}{\square\square} = \begin{matrix}\square\square\square\\\square\square\square\end{matrix} \times \square\square$$

$$k = \begin{matrix}\square\square\square\\\square\square\square\\\square\square\square\\\square\square\square\end{matrix} \qquad k = 12$$

Family Letter

2B Equations with Rational Numbers *continued*

The easiest two-step equations to solve are those that have only one variable term. The student will learn to solve these types of equations by performing inverse (opposite) operations to isolate the variable.

Solve. $\mathbf{-4x + 8 = 16}$

$$\begin{array}{r} -4x + 8 = 16 \\ \underline{\quad -8 \quad -8} \\ -4x = 8 \end{array}$$

Get the x by itself.
First subtract 8 from both sides.

$$\frac{-4x}{-4} = \frac{8}{-4}$$

Next, divide both sides by –4.

$$x = -2$$

Check your answer by substituting –2 for x.

$-4(-2) + 8 = 8 + 8 = 16$ ✔

If an equation has a variable in the numerator of a fraction, the student will need to "clear" the fraction by multiplying each term by the denominator of the fraction. This will make the problem less complicated to solve.

Solve. $\mathbf{\frac{c+5}{3} = 7}$

$$3 \bullet \frac{c+5}{3} = 7 \bullet 3$$

Multiply both sides of the equation by 3.

$$c + 5 = 21$$

$$\begin{array}{r} c + 5 = 21 \\ \underline{\quad -5 \quad -5} \\ c = 16 \end{array}$$

Subtract 5 from both sides.

Solving equations to find the value of a variable is an invaluable skill in the student's mathematical education. Be sure to practice solving a variety of one- and two-step equations with the student.

Sincerely,

Name ______________________ Date ______________ Class ______________

CHAPTER 2

At-Home Practice

2B Equations with Rational Numbers

Evaluate each expression for the given value of the variable.

1. $\frac{3.1}{x}$ for $x = 0.2$

2. $x \div \frac{7}{9}$ for $x = 2\frac{4}{9}$

3. $\frac{5.4}{x}$ for $x = 1.5$

Solve.

4. $x - 18.5 = -32$

5. $\frac{x}{11.4} = 3$

6. $x - 3.62 = 7.18$

7. $\frac{3}{10}x = \frac{4}{15}$

8. $x - \frac{1}{3} = \frac{3}{5}$

9. $\frac{x}{9} = \frac{2}{3}$

Solve. Check each answer.

10. $4x - 7 = 17$

11. $\frac{h}{6} - 11 = -14$

12. $\frac{y + 5}{6} = 4$

13. $19 = 7 - 3x$

14. $\frac{3 - 6k}{3} = -9$

15. $5 + \frac{m}{3} = 1$

Answers: 1. 15.5 **2.** $3\frac{1}{7}$ **3.** 3.6 **4.** $x = -13.5$ **5.** $x = 34.2$ **6.** $x = 10.8$ **7.** $x = \frac{8}{9}$ **8.** $x = \frac{14}{15}$ **9.** $x = 6$ **10.** $x = 6$ **11.** $h = -18$ **12.** $y = 19$ **13.** $x = -4$ **14.** $k = 5$ **15.** $m = -12$

Name ______________________ Date ______________ Class ______________

CHAPTER 2

Family Fun

Equation Riddle

Solve the equations. Fill in each blank with the variable equal to the number below it to answer the riddle.

1. $4o - 0.75 = 6.75$
2. $\frac{t}{5} - 7 = -12$
3. $5p + p = 54$
4. $\frac{x}{10} - \frac{15}{10} = 4\frac{1}{2}$
5. $3s + 17 = 56$
6. $4e + 0.625 = 3.125$
7. $7w + 54 = 61$
8. $40 + 3a = -5$

What dance do equations do?

___	___	___	___	___		___	___	___	-	___	___	___	___
–25	0.625	60	–15	13		–25	1	1.875		13	–25	0.625	9

Answer: texas two-step

Name ____________________ Date ____________ Class ____________

LESSON 2-7

Practice A

Solving Equations with Rational Numbers

Solve.

1. $x + 1.2 = 4.6$

2. $a - 3.4 = 5$

3. $2.2m = 4.4$

4. $\frac{x}{1.3} = 2$

5. $6.7 + w = -1.1$

6. $\frac{n}{1.9} = -3.8$

7. $7.2 = -0.9y$

8. $k - 4.05 = 6.2$

9. $\frac{d}{-3.2} = -3.75$

10. $-\frac{2}{5} + x = \frac{2}{5}$

11. $\frac{1}{4}x = \frac{1}{2}$

12. $\frac{1}{3}a = \frac{3}{4}$

13. $x - \frac{3}{2} = \frac{1}{5}$

14. $x - \frac{3}{7} = -\frac{5}{7}$

15. $-\frac{5}{6}a = \frac{5}{8}$

16. Elisa can reach $77\frac{3}{4}$ in. high. The ceiling is $90\frac{1}{2}$ in. high. How much higher is the ceiling than Elisa's highest reach?

17. Nolan Makes $10.60 an hour at his after-school job. Last week he worked 11.25 hr. How much was Nolan paid for the week?

Name ______________________ Date ______________ Class ______________

LESSON 2-7

Practice B

Solving Equations with Rational Numbers

Solve.

1. $x + 6.8 = 12.19$ ________

2. $y - 10.24 = 5.3$ ________

3. $0.05w = 6.25$ ________

4. $\frac{a}{9.05} = 8.2$ ________

5. $-12.41 + x = -0.06$ ________

6. $\frac{d}{-8.4} = -10.2$ ________

7. $-2.89 = 1.7m$ ________

8. $n - 8.09 = -11.65$ ________

9. $\frac{x}{5.4} = -7.18$ ________

10. $\frac{7}{9} + x = 1\frac{1}{9}$ ________

11. $\frac{6}{11}y = -\frac{18}{22}$ ________

12. $\frac{7}{10}d = \frac{21}{20}$ ________

13. $x - \left(-\frac{9}{14}\right) = \frac{5}{7}$ ________

14. $x - \frac{15}{21} = 2\frac{6}{7}$ ________

15. $-\frac{8}{15}a = \frac{9}{10}$ ________

16. A recipe calls for $2\frac{1}{3}$ cups of flour and $1\frac{1}{4}$ cups of sugar. If the recipe is tripled, how much flour and sugar will be needed?

____

17. Daniel filled the gas tank in his car with 14.6 gal of gas. He then drove 284.7 mi before needing to fill up his tank with gas again. How many miles did the car get to a gallon of gasoline?

____

Name ______________________ Date ______________ Class ______________

LESSON 2-7

Practice C

Solving Equations with Rational Numbers

Solve.

1. $x + 102.8 = 89.06$

2. $62.5m = 2587.5$

3. $\frac{w}{38.7} = 51.06$

______________ ______________ ______________

4. $-10\frac{5}{18} + x = -12\frac{3}{10}$

5. $5\frac{2}{15}a = 3\frac{2}{25}$

6. $y - \left(-6\frac{1}{16}\right) = 11\frac{3}{40}$

______________ ______________ ______________

7. A photo that is $3\frac{1}{2}$ in. by $5\frac{1}{4}$ in. is enlarged to three times its original size. What are the new dimensions of the photo?

____

8. Mars has two very small elliptical shaped moons Deimos and Phobus. They were discovered in August 1877, by Asaph Hall, an American Astronomer. The inner satellite is Phobos. It is 16.78 mi long. Deimos is the outer moon and is 9.32 mi long. What is the difference in the length of the two moons?

____

9. Mr. Crowley bought lunch for himself and eight of his employees. Each had a sandwich platter that costs $5.85 and a drink that costs $1.25. Five of the employees had dessert that costs $1.50. Mr. Crowley gave the delivery person $80 and told her to keep the change as a tip. How much was the delivery person's tip?

____

10. Two of the greatest rainfalls ever recorded were on July 4, 1956. In Unionville, Maryland it rained 1.23 in. in 1 min. In Curtea-de-Arges, Romania on July 7, 1889 it rained 8.1 in. in 20 min. If it had rained for 20 min in Unionville at its same record pace, what would be the difference between the two rainfall amounts?

____

Name ______________________ Date ______________ Class ______________

LESSON 2-7

Review for Mastery

Solving Equations with Rational Numbers

Solving equations with rational numbers is basically the same as solving equations with integers or whole numbers:

Use inverse operations to isolate the variable.

$\frac{1}{4}z = -16$

$4 \cdot \frac{1}{4}z = -16 \cdot 4$

$z = -64$

Multiply each side by 4.

$y - \frac{3}{8} = \frac{7}{8}$

$+\frac{3}{8} \quad +\frac{3}{8}$

$y = \frac{10}{8} = 1\frac{2}{8} = 1\frac{1}{4}$

Add $\frac{3}{8}$ to each side.

$x + 3.5 = -17.42$

$- 3.5 \quad - 3.5$

$x = -20.92$

Subtract 3.5 from each side.

$-26t = 317.2$

$\frac{-26t}{-26} = \frac{317.2}{-26}$

$t = -12.2$

Divide each side by −26.

Tell what you would do to isolate the variable.

1. $x - 1.4 = 7.82$ ______________
2. $\frac{1}{4} + y = \frac{7}{4}$ ______________
3. $3z = 5$ ______________

Solve each equation.

4. $14x = -129.5$ ______________
5. $\frac{1}{3}y = 27$ ______________
6. $265.2 = \frac{z}{22.1}$ ______________
7. $x + 53.8 = -1.2$ ______________
8. $25 = \frac{1}{5}k$ ______________
9. $m - \frac{2}{3} = \frac{3}{5}$ ______________

Name ______________________ Date ______________ Class ______________

LESSON 2-7

Challenge

Location, Location, Location

An equation that has a variable in the denominator of one or more of its terms is called a **fractional equation.**

One method of solution is to clear the equation of fractions by multiplying each side of the equation by the LCD.

$\frac{1}{2} + \frac{1}{x} = \frac{3}{5}$ The LCD of 2, x, and 5 is $10x$, with $x \neq 0$.

$10x\left(\frac{1}{2} + \frac{1}{x}\right) = 10x\left(\frac{3}{5}\right)$ Multiply each side by $10x$.

$10x \cdot \frac{1}{2} + 10x \cdot \frac{1}{x} = 10x \cdot \frac{3}{5}$ Distributive Property

$5x + 10 = 6x$ Simplify.

$5x - 5x + 10 = 6x - 5x$ Subtract $5x$ from each side.

$10 = x$

Check:

$\frac{1}{2} + \frac{1}{x} = \frac{3}{5}$

$\frac{1}{2} + \frac{1}{10} \stackrel{?}{=} \frac{3}{5}$ Substitute 10 for x in the original equation.

$\frac{5}{10} + \frac{1}{10} \stackrel{?}{=} \frac{3}{5}$ Do not repeat the method of solution.

$\frac{6}{10} \stackrel{?}{=} \frac{3}{5}$ Work each side separately.

$\frac{3}{5} = \frac{3}{5}$ ✔

Solve and check.

1. $\frac{4}{7} + \frac{2}{x} = \frac{2}{3}$

2. $\frac{10}{x} + \frac{8}{x} = 9$

3. $\frac{15}{x} = 7 + \frac{9}{2x}$

________________ ________________ ________________

Name ______________________ Date ______________ Class ______________

LESSON 2-7

Problem Solving

Solving Equations with Rational Numbers

Write the correct answer.

1. In the last 150 years, the average height of people in industrialized nations has increased by $\frac{1}{3}$ foot. Today, American men have an average height of $5\frac{7}{12}$ feet. What was the average height of American men 150 years ago?

   ____

2. Jaime has a length of ribbon that is $23\frac{1}{2}$ in. long. If she plans to cut the ribbon into pieces that are $\frac{3}{4}$ in. long, into how many pieces can she cut the ribbon? (She cannot use partial pieces.)

   ____

3. Todd's restaurant bill for dinner was \$15.55. After he left a tip, he spent a total of \$18.00 on dinner. How much money did Todd leave for a tip?

   ____

4. The difference between the boiling point and melting point of Hydrogen is 6.47 °C. The melting point of Hydrogen is –259.34 °C. What is the boiling point of Hydrogen?

   ____

Choose the letter for the best answer.

5. In 2005, a sprinter won the gold medal in the 100-m dash in with a time of 9.85 seconds. His time was 0.95 seconds faster than the winner in the 100-m dash in 1900. What was winner's time in 1900?

   A 8.95 seconds
   B 10.65 seconds
   C 10.80 seconds
   D 11.20 seconds

6. The balance in Susan's checking account was \$245.35. After the bank deposited interest into the account, her balance went to \$248.02. How much interest did the bank pay Susan?

   F \$1.01
   G \$2.67
   H \$3.95
   J \$493.37

7. After a morning shower, there was $\frac{17}{100}$ in. of rain in the rain gauge. It rained again an hour later and the rain gauge showed $\frac{1}{4}$ in. of rain. How much did it rain the second time?

   A $\frac{2}{25}$ in.　　C $\frac{21}{50}$ in.
   B $\frac{1}{6}$ in.　　D $\frac{3}{8}$ in.

8. Two-third of John's savings account is being saved for his college education. If \$2500 of his savings is for his college education, how much money in total is in his savings account?

   F \$1666.67　　H \$4250.83
   G \$3750　　J \$5000

Name ______________________ Date ______________ Class ______________

LESSON 2-7

Reading Strategies

Follow a Procedure

The rules for solving equations with rational numbers are the same as equations with whole numbers.

Get the variable by itself.	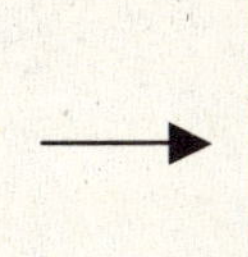	Perform the same operation on both sides to keep the equation balanced.	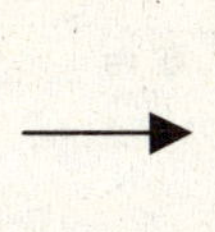	Use the rules for computing rational numbers.

Follow the steps above to help you solve $\frac{1}{4} + y = \frac{3}{4}$.

1. What is the first step to solve this equation?

____

2. What operation should you use?

____

3. Write an equation to show the subtraction of $\frac{1}{4}$ on both sides.

____

4. What is the value of *y*?

____

Follow the steps above to solve $x - 4.5 = 13$.

5. What is the first step to solve this equation?

____

6. What operation should you use?

____

7. Write an equation to show the addition of 4.5 to both sides.

____

8. Find the value of x.

____

Name ______________________ Date ______________ Class ______________

LESSON 2-7

Puzzles, Twisters & Teasers

Math Book Issues!

Solve the equations. Then use the letters of the variables to answer the riddle.

1. $s - \frac{5}{9} = \frac{1}{9}$ $\quad s =$ ______________
2. $t + \frac{1}{3} = \frac{3}{4}$ $\quad t =$ ______________
3. $m - \frac{1}{12} = \frac{5}{12}$ $\quad m =$ ______________
4. $o + \frac{5}{9} = -\frac{1}{9}$ $\quad o =$ ______________
5. $e - 17.9 = 36.8$ $\quad e =$ ______________
6. $a - 2.1 = -4.5$ $\quad a =$ ______________
7. $l + \frac{4}{13} = \frac{12}{39}$ $\quad l =$ ______________
8. $0.04n = 0.252$ $\quad n =$ ______________
9. $b \div 3.2 = -6$ $\quad b =$ ______________
10. $\frac{5}{9} + y = \frac{6}{18}$ $\quad y =$ ______________
11. $r + \frac{5}{8} = -2\frac{3}{8}$ $\quad r =$ ______________
12. $p + 3.8 = -1.6$ $\quad p =$ ______________

Why did the student return her math book?

It had ____ ____ ____ ____ ____ ____ ____

$\frac{5}{12}$ $\quad -\frac{2}{3}$ $\quad -\frac{2}{3}$ $\qquad \frac{1}{2}$ $\quad -2.4$ $\quad 6.3$ $\quad -\frac{2}{9}$

____ ____ ____ ____ ____ ____ ____ ____

-5.4 $\quad -3$ $\quad -\frac{2}{3}$ $\quad -19.2$ $\quad 0$ $\quad 54.7$ $\quad \frac{1}{2}$ $\quad \frac{2}{3}$

Name ______________________ Date ______________ Class ______________

LESSON 2-8

Practice A

Solving Two-Step Equations

Describe the operation performed on both sides of the equation in steps 2 and 4.

1. $3x + 2 = 11$

$3x + 2 - 2 = 11 - 2$ ______________

$3x = 9$

$\frac{3x}{3} = \frac{9}{3}$ ______________

$x = 3$

2. $\frac{x}{4} - 1 = -2$

$\frac{x}{4} - 1 + 1 = -2 + 1$ ______________

$\frac{x}{4} = -1$

$4\left(\frac{x}{4}\right) = 4(-1)$ ______________

$x = -4$

Solve.

3. $2x + 3 = 9$ ______________

4. $\frac{x}{3} - 1 = 5$ ______________

5. $-3a + 4 = 7$ ______________

6. $\frac{x+2}{2} = -3$ ______________

7. $5y - 2 = 28$ ______________

8. $2x - 7 = 7$ ______________

9. $\frac{w-2}{5} = -1$ ______________

10. $2r + 1 = -1$ ______________

Write and solve a two-step equation to answer the question.

11. Pearson rented a moving van for 1 day. The total rental charge is $66.00. A daily rental costs $45.00 plus $0.25 per mile. How many miles did he drive the van?

Name ______________________ Date ______________ Class ______________

LESSON 2-8

Practice B

Solving Two-Step Equations

1. The school purchased baseball equipment and uniforms for a total cost of $1762. The equipment costs $598 and the uniforms were $24.25 each. How many uniforms did the school purchase?

2. Carla runs 4 miles every day. She jogs from home to the school track, which is $\frac{3}{4}$ mile away. She then runs laps around the $\frac{1}{4}$-mile track. Carla then jogs home. How many laps does she run at the school?

Solve.

3. $\frac{a+5}{3} = 12$

4. $\frac{x+2}{4} = -2$

5. $\frac{y-4}{6} = -3$

6. $\frac{k+1}{8} = 7$

7. $0.5x - 6 = -4$

8. $\frac{x}{2} + 3 = -4$

9. $\frac{1}{5}n + 3 = 6$

10. $2a - 7 = -9$

11. $\frac{3x-1}{4} = 2$

12. $-7.8 = 4.4 + 2r$

13. $\frac{-4w+5}{-3} = -7$

14. $1.3 - 5r = 7.4$

15. A phone call costs $0.58 for the first 3 minutes and $0.15 for each additional minute. If the total charge for the call was $4.78, how many minutes was the call?

16. Seventeen less than four times a number is twenty-seven. Find the number.

Name ______________________ Date ______________ Class ______________

LESSON 2-8

Practice C

Solving Two-Step Equations

Write an equation for each sentence, then solve.

1. A number multiplied by five and increased by 3 is 28.

____

2. Eighteen decreased by 4 times a number is 62.

____

3. The sum of 3 times a number and 7, divided by 5, is 17.

____

4. The quotient of a number and 5, minus 2, is 8.

____

Solve.

5. $-13 = -3x + 14$

6. $2w - 5 = 4$

7. $\dfrac{x - 5}{3} = -10$

8. $\dfrac{2}{3}n - 7 = 19$

9. $1.4x + 0.8 = -1.3$

10. $\dfrac{1}{4}y + \dfrac{3}{4} = 2$

11. $15n - 62 = -17$

12. $\dfrac{3}{8}a - 4 = -\dfrac{1}{4}$

13. $\dfrac{2d + 9}{6} = 11$

14. $24.5 = 16.1 - 2.4r$

15. $\dfrac{7x - 2}{6} = -5$

16. $\dfrac{5}{6} - \dfrac{a}{4} = \dfrac{1}{3}$

17. Larissa is planning for a trip that cost $2145. She has $952.50 saved and is going to set aside $\dfrac{1}{2}$ of her weekly salary from her part-time job. Larissa earns $265 a week. How many weeks will it take her to earn the rest of the money needed for the trip? ______________

18. Noel bought a printer for $10 less than half its original price. If Noel paid $88 for the printer, what was the original price? ______________

Name ______________________ Date ______________ Class ______________

LESSON 2-8

Review for Mastery

Solving Two-Step Equations

To solve an equation, it is important to first note how it is formed.

Then, work backward to undo each operation.

$4z + 3 = 15$	$\frac{z}{4} - 3 = 7$	$\frac{z+3}{4} = 7$
The variable is multiplied by 4 and then 3 is added.	The variable is divided by 4 and then 3 is subtracted.	3 is added to the variable and then the result is divided by 4.
To solve, first subtract 3 and then divide by 4.	To solve, first add 3 and then multiply by 4.	To solve, multiply by 4 and then subtract 3.

Describe how each equation is formed.
Then, tell the steps needed to solve.

1. $3x - 5 = 7$

   The variable is ______________________ and then ______________________.

   To solve, first ______________________ and then ______________________.

2. $\frac{x}{3} + 5 = 7$

   The variable is ______________________ and then ______________________.

   To solve, first ______________________ and then ______________________.

3. $\frac{x+5}{3} = 7$

   5 is ______________________ and then the result is ______________________.

   To solve, first ______________________ and then ______________________.

4. $10 = -3x - 2$

   The variable is ______________________ and then ______________________.

   To solve, first ______________________ and then ______________________.

5. $10 = \frac{x-2}{5}$

   2 is ______________________ the variable and then the result is ______________.

   To solve, first ______________________ and then ______________________.

Name ______________________ Date ______________ Class ______________

LESSON 2-8

Review for Mastery

Solving Two-Step Equations (continued)

To isolate the variable, work backward using inverse operations.

The variable is multiplied by 2 and then 3 is added.

$2x + 3 = 11$ To undo addition,

$\underline{\quad -3 \quad -3}$ subtract 3.

$2x = 8$ To undo multiplication,

$\frac{2x}{2} = \frac{8}{2}$ divide by 2.

$x = 4$

Check: Substitute 4 for *x*.

$2(4) + 3 \stackrel{?}{=} 11$

$8 + 3 \stackrel{?}{=} 11$

$11 = 11$ ✓

The variable is divided by 2 and then 3 is subtracted.

$\frac{x}{2} - 3 = 11$ To undo subtraction,

$\underline{\quad +3 \quad +3}$ add 3.

$\frac{x}{2} = 14$ To undo division,

$2 \cdot \frac{x}{2} = 2 \cdot 14$ multiply by 2.

$x = 28$

Check: Substitute 28 for *x*.

$\frac{28}{2} - 3 \stackrel{?}{=} 11$

$14 - 3 \stackrel{?}{=} 11$

$11 = 11$ ✓

Complete to solve and check each equation.

6. $3t + 7 = 19$ To undo addition, subtract.

____ ____

$3t =$ ____ To undo multiplication, divide.

$3t \div$ ____ $=$ ____ $\div$ ____

$t =$ ____

Check: $3t + 7 = 19$

$3($____$) + 7 \stackrel{?}{=} 19$ Substitute for *t*.

____ $+ 7 \stackrel{?}{=} 19$

____

7. $\frac{w}{3} - 7 = 5$ To undo subtraction, add.

____ ____

$\frac{w}{3} =$ ____ To undo division, multiply.

____ $\cdot \frac{w}{3} =$ ____ $\cdot 12$

$w =$ ____

Check: $\frac{w}{3} - 7 = 5$

$\frac{\quad}{3} - 7 \stackrel{?}{=} 5$ Substitute.

____ $- 7 \stackrel{?}{=} 5$

____

8. $\frac{z-3}{2} = 8$ To undo division, multiply.

____ $\cdot \frac{z-3}{2} =$ ____ $\cdot 8$

$z - 3 =$ ____ To undo subtraction, add.

____ $=$ ____

$z =$ ____

Check: $\frac{z-3}{2} = 8$

$\frac{\quad - 3}{2} \stackrel{?}{=} 8$ Substitute.

$\frac{\quad}{2} \stackrel{?}{=} 8$

____

Name ________________________ Date ______________ Class ______________

LESSON 2-8

Challenge

Work It Algebraically!

An equation may be used to solve a problem involving probability.

A bag contains marbles of four colors: red, white, blue, and yellow. There are 3 more blue marbles than red, and 48 marbles in all. How many blue marbles are there if, in one draw, the probability of getting a blue marble is $\frac{5}{12}$?

Let x = the number of blue marbles.

$$P(\text{blue}) = \frac{\text{number of successes}}{\text{total number}}$$

$$\frac{5}{12} = \frac{x}{48}$$

$$\frac{5}{12} \cdot 48 = 48 \cdot \frac{x}{48} \qquad \text{Multiply by 48.}$$

$$20 = x$$

So, there are 20 blue marbles in the bag.

Write and solve an equation for each problem.

1. A box has four kinds of candies: lime, orange, cherry, and mint. There are 9 more lime candies than orange, and 36 candies in all. How many lime candies are there if, in one draw, the probability of getting a lime candy is $\frac{4}{9}$?

   There are _____ lime candies.

2. A carton has four kinds of cookies: lemon, mint, vanilla, and chocolate. There are 7 fewer mint cookies than lemon, and 64 cookies in all. How many mint cookies are there if, in one draw, the probability of getting a lemon cookie is $\frac{5}{8}$?

   There are _____ mint cookies.

Name ______________________ Date ______________ Class ______________

LESSON 2-8

Problem Solving

Solving Two-Step Equations

The chart below describes three different long-distance calling plans. Jamie has budgeted $20 per month for long-distance calls. Write the correct answer.

Plan	Monthly Access Fee	Charge per minute
A	$3.95	$0.08
B	$8.95	$0.06
C	$0	$0.10

1. How many minutes will Jamie be able to use per month with plan A? Round to the nearest minute.

____

2. How many minutes will Jamie be able to use per month with plan B? Round to the nearest minute.

____

3. How many minutes will Jamie be able to use per month with plan C? Round to the nearest minute.

____

4. Which plan is the best deal for Jamie's budget?

____

5. Nolan has budgeted $50 per month for long distance. Which plan is the best deal for Nolan's budget?

____

The table describes four different car loans that Susana can get to finance her new car. The total column gives the amount she will end up paying for the car including the down payment and the payments with interest. Choose the letter for the best answer.

Loan	Down Payment	Number of Months	Total
A	$2000	60	$19,821.20
B	$1000	48	$19,390.72
C	$0	60	$20,197.20

6. How much will Susana pay each month with loan A?

A $252.04 C $330.35
B $297.02 D $353.68

7. How much will Susana pay each month with loan B?

F $300.85 H $323.17
G $306.50 J $383.14

8. How much will Susana pay each month with loan C?

A $336.62 C $369.95
B $352.28 D $420.78

9. Which loan will give Susana the smallest monthly payment?

F Loan A H Loan C
G Loan B J They are equal

Name ______________________ Date ______________ Class ______________

LESSON 2-8

Reading Strategies

Analyze Information

Break a problem into parts and analyze the information.

Jill has $8 in her pocket now. She had $20 when she left for the movies. How much money did she spend?

Answer the questions in Exercises 1–4 to solve this problem.

1. How much money did Jill start with?

____

2. How much money does Jill have left?

____

3. What is the difference between these two amounts?

____

4. How much money did Jill spend?

____

Mark paid $45 at the music store for 3 CDs and a pack of batteries, before tax. The batteries cost $6. How much did Mark pay for each of the CDs?

Answer the questions in Exercises 5–9 to solve this problem.

5. How much did Mark spend at the music store?

____

6. How much did Mark spend on batteries?

____

7. What is the difference between these two amounts?

____

8. Since Mark paid $39 for CDs, divide $39 by 3.

____

9. How much did Mark pay for each CD?

____

Name ______________________ Date ______________ Class ______________

LESSON 2-8

Puzzles, Twisters & Teasers

Have a Ball!

Solve the equations. Match the letters of the variables to the answers to solve the riddle.

1. $7 + \frac{l}{5} = -4$ **l =** ____________
2. $46 - 3t = -23$ **t =** ____________
3. $8 = 6 + \frac{a}{4}$ **a =** ____________
4. $6h + 24 = 0$ **h =** ____________
5. $9 = -5b - 23$ **b =** ____________
6. $15 - 3e = -6$ **e =** ____________
7. $15w - 4 = 41$ **w =** ____________
8. $6s + 3 = -27$ **s =** ____________
9. $14o - 17 = 39$ **o =** ____________
10. $\frac{n}{-3} - 2 = 8$ **n =** ____________

Where do penguins go to dance?

A $\underset{23}{\_\_\_}$ **T** $\underset{-4}{\_\_\_}$ $\underset{7}{\_\_\_}$ $\underset{-5}{\_\_\_}$ $\underset{-30}{\_\_\_}$ $\underset{4}{\_\_\_}$ $\underset{3}{\_\_\_}$ $\underset{-6.4}{\_\_\_}$ $\underset{8}{\_\_\_}$ **L** $\underset{-55}{\_\_\_}$

Answers

LESSON 2-1

Practice A

1. $\frac{1}{3}$
2. $\frac{1}{3}$
3. $-\frac{1}{4}$
4. $\frac{1}{4}$
5. $\frac{7}{12}$
6. $-\frac{3}{7}$
7. $\frac{10}{21}$
8. $-\frac{4}{9}$
9. $\frac{2}{5}$
10. $-\frac{7}{20}$
11. $\frac{21}{200}$
12. $1\frac{1}{5}$
13. $-\frac{17}{20}$
14. $\frac{13}{40}$
15. $\frac{1}{500}$
16. $2\frac{3}{10}$
17. $\frac{7}{25}$
18. $-1\frac{1}{4}$
19. $\frac{8}{125}$
20. $\frac{3}{400}$
21. $0.\overline{1}$
22. 0.5625
23. –0.55
24. 1.2
25. $0.1\overline{3}$
26. $-2.58\overline{3}$
27. 0.03
28. 5.32
29. Possible answer: $\frac{5}{12}$

Practice B

1. $\frac{2}{3}$
2. $\frac{1}{2}$
3. $\frac{1}{4}$
4. $-\frac{1}{4}$
5. $\frac{3}{8}$
6. $-\frac{1}{12}$
7. $-\frac{2}{9}$
8. $\frac{1}{6}$
9. $\frac{18}{25}$
10. $\frac{29}{500}$
11. $-1\frac{13}{20}$
12. $2\frac{1}{10}$
13. $\frac{9}{250}$
14. $-4\frac{3}{50}$
15. $2\frac{61}{200}$
16. $\frac{4}{625}$
17. $-\frac{3}{5}$
18. $6\frac{19}{20}$
19. $\frac{2}{125}$
20. $\frac{1}{2000}$
21. 0.125
22. $2.\overline{6}$
23. $0.9\overline{3}$
24. 3.2
25. 0.6875
26. $0.\overline{7}$
27. 0.8
28. 1.24
29. Possible answer: $\frac{5}{24}$

Practice C

1. $\frac{9}{10}$
2. $2\frac{1}{2}$
3. $-\frac{9}{25}$
4. $-\frac{43}{200}$
5. $-4\frac{1}{50}$
6. $\frac{17}{2000}$
7. $1\frac{3}{500}$
8. $\frac{9}{20}$
9. 0.6
10. –0.44
11. 2.8125
12. –0.2
13. 0.1875
14. $0.1\overline{6}$

15. $1.58\overline{3}$

16. 1.95

17. Possible answer: $\frac{5}{48}$

18. a. $\frac{1}{9}$; a. $\frac{1}{8}$; a. $\frac{2}{5}$

b. $\frac{1}{3 \cdot 3}$; b. $\frac{1}{2 \cdot 2 \cdot 2}$; b. $\frac{2}{5}$

c. $0.11\overline{1}$; repeating

c. 0.125; terminating

c. 0.4; terminating

Review for Mastery

1. $\frac{1}{2}$
2. $\frac{15 \div 15}{45 \div 15}$; $\frac{1}{3}$
3. $\frac{12 \div 6}{30 \div 6}$; $\frac{2}{5}$
4. $\frac{12 \div 12}{24 \div 12}$; $\frac{1}{2}$
5. $\frac{5 \div 5}{35 \div 5}$; $\frac{1}{7}$
6. $\frac{14 \div 7}{49 \div 7}$; $\frac{2}{7}$
7. $\frac{1}{7}$
8. $\frac{3}{10}$
9. $\frac{2}{9}$
10. $\frac{1}{4}$
11. $\frac{3}{8}$
12. $\frac{11}{20}$
13. $\frac{8}{25}$
14. 3.75
15. $0.8\overline{3}$
16. $3.\overline{6}$
17. 2.5
18. 1.875
19. $4.\overline{6}$
20. 5.5
21. $5.1\overline{6}$
22. 10.5

Challenge

1. $0.\overline{1}$
2. $0.\overline{2}$
3. $0.\overline{3}$
4. $0.\overline{4}$
5. $0.\overline{6}$
6. $0.\overline{8}$
7. $\frac{5}{9}$
8. $\frac{7}{9}$
9. $\frac{9}{9} = 1$
10. $0.\overline{42}$
11. $0.\overline{358}$
12. $0.\overline{4276}$
13. $0.\overline{76}$
14. $0.\overline{732}$
15. $0.\overline{1957}$
16. $\frac{45}{99}$
17. $\frac{148}{999}$
18. $\frac{7213}{9999}$
19. A single digit repeating decimal equates to a fraction with denominator 9, a 2–digit repeating decimal to denominator 99, and so on.

Problem Solving

1.

Decimal	Decimal	Decimal
0.25	0.171875	0.09375
0.234375	0.15625	0.078125
0.21875	0.140625	0.0625
0.203125	0.125	
0.1875	0.109375	

2. Terminating decimals
3. D
4. H
5. D
6. F
7. A

Reading Strategies

1. any number that can be written in the form $\frac{a}{b}$
2. yes, because it can be written as $\frac{62}{100}$
3. yes, because it can be written as $\frac{7}{3}$
4. no, because it cannot be written as a decimal that terminates or repeats.
5. yes, because it can be written as $\frac{-8}{1}$
6. yes, because it can be written $\frac{0}{1}$

Puzzles, Twisters & Teasers

H O L E Q U I V A L E N T N C
M I R U J I N O N Z E R O U Y
O P R F A C T O R X D E B M F
Q W A U I O E W E R T Y L E I
O T T L P J G O K I Y T R R L
B Y I M J U E W E R T Y U A P
N J O A S P R I M E T Y U T M
D E N O M I N A T O R I L O I
X W A R E L A T I V E L Y R S
P L L Q A Z X S W E D C V F R

H O L E

LESSON 2-2

Practice A

1. 21 > 20; >
2. 16 < 15; <
3. > 0.25; >
4. $\frac{1}{6} < \frac{2}{6}$; <
5. 0.09 < 0.5; <
6. −0.6 = −0.6; =
7. >
8. >
9. <
10. <
11. >
12. <
13. >
14. <
15. >
16. $\frac{1}{4}$ mile, $1\frac{9}{10}$ miles, 2.05 miles, $2\frac{2}{5}$ miles

Practice B

1. >
2. <
3. >
4. >
5. >
6. <
7. <
8. =
9. <
10. =
11. <
12. <
13. <
14. <
15. =
16. <
17. >
18. =
19. −1.8 minutes, $-1\frac{2}{3}$ minutes, −1.45 minutes, $-1\frac{3}{8}$ minutes
20. Possible answer: 3.2 miles

Practice C

1. >
2. <
3. >
4. =
5. >
6. >
7. <
8. =
9. <
10. <
11. <
12. >

Possible answers are given for Ex. 13–18.

13. 0.18
14. 0.701
15. 0
16. $-\frac{1}{3}$
17. 1.455
18. 0.7
19. $\frac{1}{4}$, 0.33, $\frac{3}{8}$, 0.4; Kyle
20. Possible answer: $2\frac{3}{4}$ miles; 2.7 miles

Review for Mastery

1. >
2. >
3. <
4. <
5. >
6. <
7. <
8. <

Challenge

1. $\frac{2}{3}$
2. $\frac{8}{9}$
3. $\frac{5}{11}$
4. $\frac{1}{11}$
5. $\frac{7}{11}$
6. $\frac{1}{6}$
7. $\frac{2}{9}$
8. $\frac{5}{6}$
9. $\frac{5}{12}$
10. $\frac{7}{12}$

Problem Solving

1. 8.5 m, 8.54 m, 8.67 m, 8.72 m
2. Point B
3. scientists aboard the submarine
4. the second heat
5. B
6. J
7. C

Reading Strategies

1. $\frac{1}{2}$
2. 1
3. $\frac{35}{67} < \frac{11}{13}$
4. 1
5. 0
6. $\frac{17}{20} > \frac{6}{35}$
7. $\frac{21}{40}$ is close to $\frac{1}{2}$ and $\frac{19}{21}$ is close to 1. $\frac{1}{2} < 1$, so $\frac{21}{40} < \frac{19}{21}$.

Puzzles, Twisters, and Teasers

1. –2.7; S
2. $-2\frac{1}{2}$; H
3. –2.05; E
4. –1.3; W
5. $-1\frac{1}{4}$; A
6. $-1\frac{1}{8}$; S
7. $-\frac{5}{6}$; R
8. –0.6; A
9. $-\frac{1}{5}$; T
10. 0; I
11. $\frac{1}{8}$; O
12. $\frac{3}{4}$; N
13. 0.8; A
14. 0.95; L
15. She was rational.

LESSON 2-3

Practice A

1. $9\frac{1}{2}$ in.
2. 0.05 sec
3. 0.4

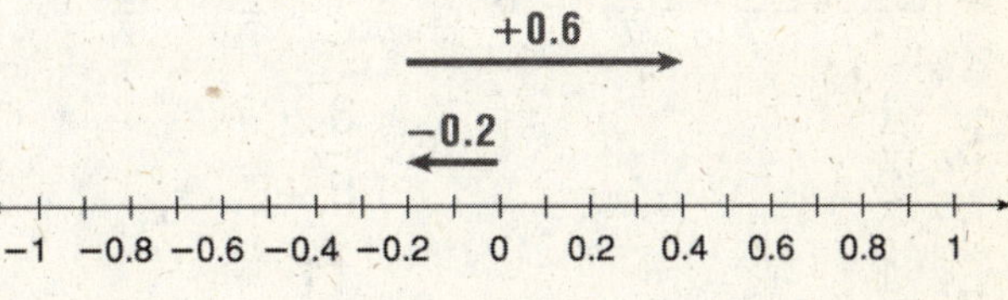

4. $\frac{4}{5}$

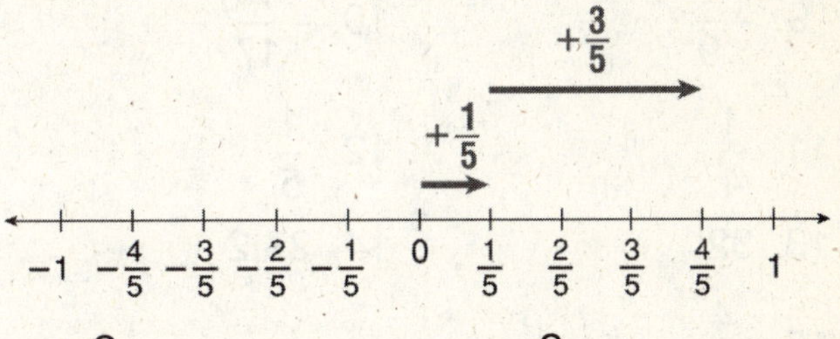

5. $\frac{2}{3}$
6. $\frac{2}{3}$
7. $\frac{1}{5}$
8. $-\frac{1}{5}$
9. $-\frac{3}{7}$
10. $-\frac{1}{3}$
11. $\frac{3}{4}$
12. 1
13. 16.7
14. 23.4
15. $\frac{4}{11}$

Practice B

1. \$4.68
2. $6\frac{1}{2}$ in.
3. –0.1

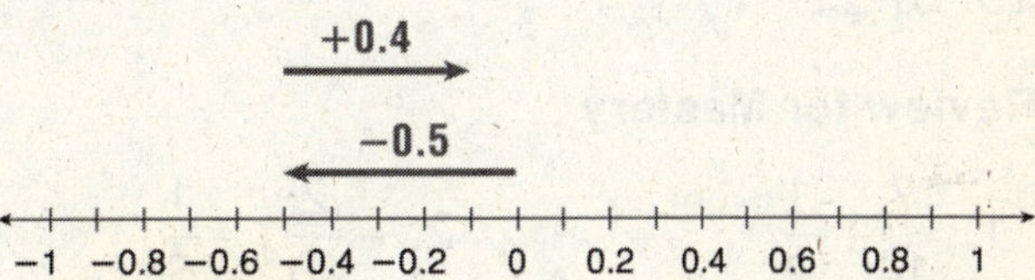

4. $\frac{4}{7}$

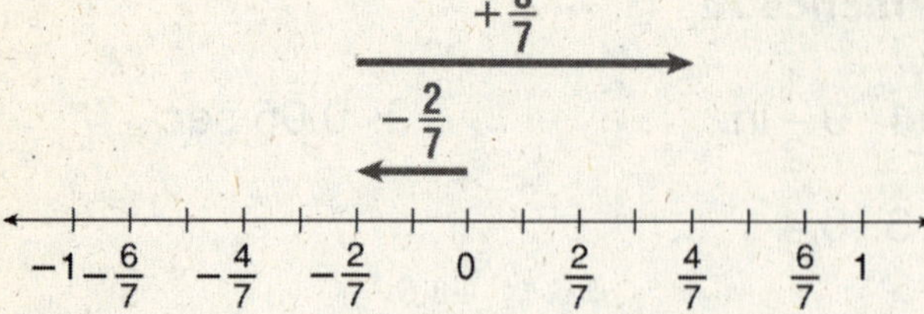

5. $\frac{1}{2}$
6. $\frac{3}{5}$
7. $\frac{1}{7}$
8. $\frac{11}{15}$
9. $-\frac{1}{9}$
10. $-\frac{10}{17}$
11. $\frac{1}{4}$
12. $\frac{1}{5}$
13. 32
14. 27.2
15. $\frac{4}{15}$

Practice C

1. $\frac{1}{8}$
2. 37.25 mi
3. 58 in.
4. $1028.44
5. $\frac{12}{17}$
6. $\frac{3}{5}$
7. $\frac{4}{5}$
8. $-\frac{1}{5}$
9. $-\frac{4}{5}$
10. $\frac{3}{4}$
11. $\frac{1}{3}$
12. $-\frac{4}{5}$
13. 104.973
14. 1
15. 37.27

Review for Mastery

1. $\frac{7}{14}$; $\frac{1}{2}$
2. $-\frac{2}{10}$; $-\frac{1}{5}$
3. $-\frac{8}{12}$; $-\frac{2}{3}$
4. $\frac{6}{9}$; $\frac{2}{3}$
5. $\frac{12}{15}$; $\frac{4}{5}$
6. $-\frac{8}{24}$; $-\frac{1}{3}$
7. 17.79
8. 52.09
9. 14.99
10. 111.17
11. 32.87
12. 0.09

Challenge

1. 4.47
2. $-\frac{1}{2}$
3. $\frac{3}{8}$
4. −0.2
5. $-1\frac{1}{4}$
6. −19.1
7. −6.4
8. $\frac{1}{5}$
9. −3
10. $\frac{3}{4}$
11. $\frac{9}{10}$
12. 1.6
13. 0
14. −1
15. 1.52
16. −38.4
17. $-\frac{7}{8}$
18. −2.05
19. 1.8

S U R E   Y O U   A R E
N O T   L E S S   T H A N
Z E R O
I' M   P O S I T I V E

Problem Solving

1. 0.18 s
2. 15.5 inches
3. $\frac{99}{100}$
4. 2 cups
5. D
6. G
7. A

Reading Strategies

1. at 0
2. to the left; because negative numbers are to the left of zero

3. 6 of the tenths
4. to the right
5. 18 of the tenths
6. 1.2
7. to the left, because negative numbers are to the left of 0
8. 4 of the sixths
9. to the right
10. 9 of the sixths
11. $\frac{5}{6}$

Puzzles, Twisters & Teasers

$-\frac{2}{3}$; 1.6; $\frac{5}{11}$; $-\frac{12}{13}$; $\frac{4}{5}$; $-\frac{1}{2}$; 0.8; $-\frac{6}{25}$; $\frac{23}{21}$; 1.2

F O R G O T   T H E
W O R D S

LESSON 2-4

Practice A

1. $\frac{5}{3}$ or $1\frac{2}{3}$
2. $-\frac{4}{5}$
3. $\frac{2}{3}$
4. $-\frac{2}{3}$
5. $-\frac{2}{7}$
6. $\frac{1}{4}$
7. $-\frac{1}{12}$
8. $\frac{1}{9}$
9. $\frac{5}{7}$
10. $-\frac{1}{12}$
11. $-\frac{3}{5}$
12. 2
13. –6
14. $\frac{15}{32}$
15. $-1\frac{3}{10}$
16. $\frac{1}{4}$
17. –16
18. 0.0204
19. 73.08
20. 14.84
21. –0.02198
22. –5.36
23. 0.6
24. –15.8
25. $22.50

Practice B

1. 6
2. –3
3. $-7\frac{1}{2}$
4. $3\frac{1}{2}$
5. $-\frac{4}{27}$
6. $\frac{7}{18}$
7. $-\frac{1}{8}$
8. $\frac{3}{22}$
9. $-\frac{3}{16}$
10. $-\frac{9}{25}$
11. $-\frac{2}{17}$
12. $-\frac{3}{10}$
13. $-8\frac{2}{3}$
14. $1\frac{1}{32}$
15. $2\frac{2}{15}$
16. $-2\frac{1}{12}$
17. 10.4
18. 0.0212
19. 27.3
20. –16.26
21. –0.0924
22. –3.9
23. 3.485
24. –50.4
25. –0.75
26. 5.168
27. –0.9
28. 0.12
29. $180

Practice C

1. 8
2. $-3\frac{1}{2}$
3. $-2\frac{1}{2}$
4. $-2\frac{1}{2}$
5. $\frac{1}{12}$
6. $\frac{1}{12}$
7. $-\frac{2}{5}$
8. $\frac{1}{21}$
9. $\frac{1}{5}$
10. $-\frac{1}{2}$
11. $-\frac{7}{30}$
12. $-\frac{1}{2}$

13. $-7\frac{1}{2}$

14. $\frac{1}{2}$

15. $-1\frac{1}{6}$

16. $\frac{13}{24}$

17. –202.5

18. 6.8472

19. 224.58

20. 139.57

21. –13.4904

22. 20.02

23. 5.0625

24. 32.016

25. $90.95

26. $51.45; $102.90

Review for Mastery

Possible models are shown.

1. 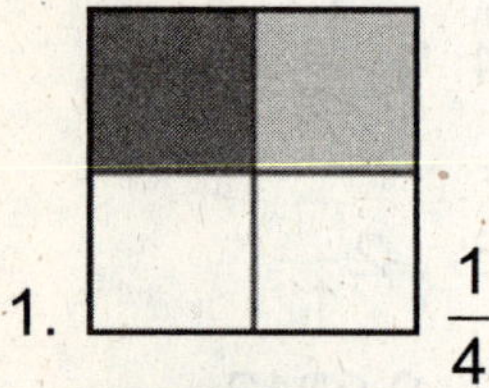 $\frac{1}{4}$

2. 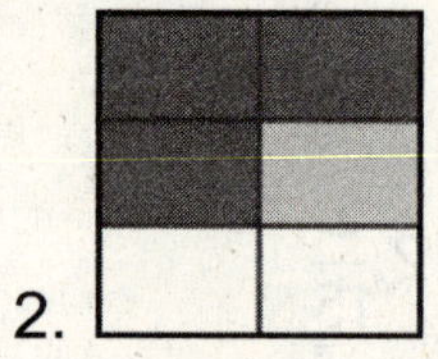 $\frac{1}{2}$

3. 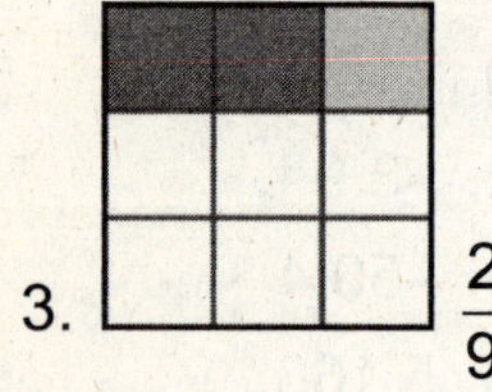 $\frac{2}{9}$

4. $\frac{2}{9}$

5. $\frac{4}{7}$

6. $\frac{9}{17}$

7. $-\frac{3}{5}$

8. $-\frac{3}{20}$

9. $1\frac{1}{2}$

Challenge

1. 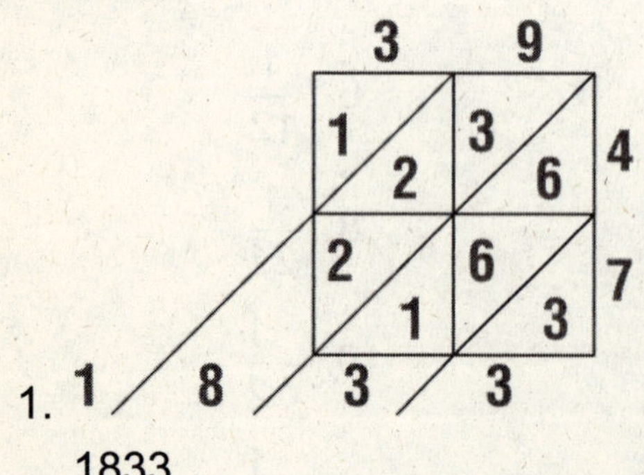

1833

2. 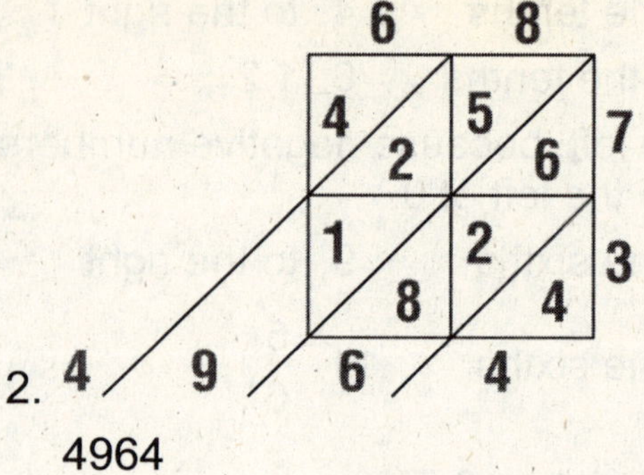

4964

3. 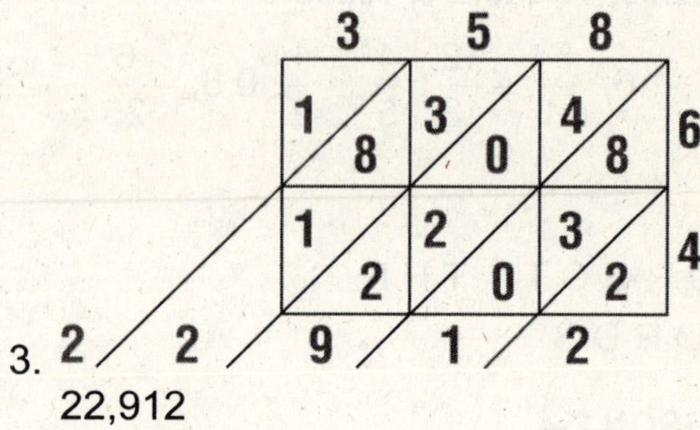

22,912

Problem Solving

1. 252 births

2. 6,120 deaths

3. 362,880 births

4. $2\frac{1}{10}$ births

5. $1\frac{1}{20}$ births

6. B

7. F

8. C

Reading Strategies

1. 6; $\frac{1}{6}$

2. $\frac{1}{6}$

3. $\frac{1}{2} \cdot \frac{1}{3} = \frac{1}{6}$

6. 8

4, 5, 7. 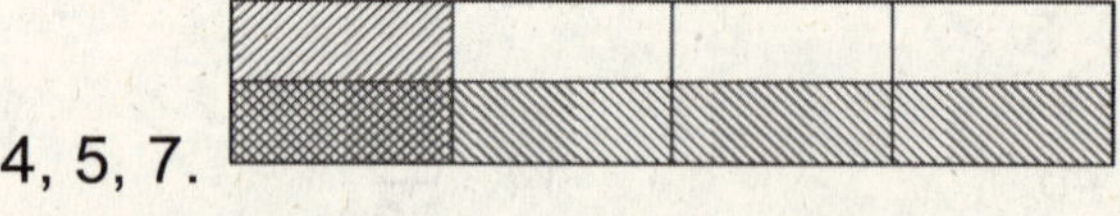

8. $\frac{1}{8}$

9. $\frac{1}{4} \cdot \frac{1}{2} = \frac{1}{8}$

Puzzles, Twisters & Teasers

$\frac{21}{40}$; $6\frac{3}{5}$; $-1\frac{1}{8}$; 4; $-8\frac{3}{4}$; $1\frac{3}{4}$; $-\frac{3}{32}$; $\frac{3}{10}$; $1\frac{7}{18}$; $\frac{4}{21}$

NEW YOLK CITY

LESSON 2-5

Practice A

1. $\frac{2}{3}$
2. $-1\frac{1}{5}$
3. $\frac{1}{2}$
4. –6
5. $\frac{1}{3}$
6. $\frac{7}{10}$
7. $-\frac{7}{10}$
8. $\frac{9}{20}$
9. $\frac{14}{15}$
10. $-\frac{1}{12}$
11. $-9\frac{1}{3}$
12. $-\frac{1}{8}$
13. 5.1
14. 25.7
15. 201
16. 13.2
17. 218
18. 5,890
19. 143
20. 2.3
21. 200
22. –45.6
23. 30.3
24. 12 books

Practice B

1. $\frac{2}{3}$
2. $-\frac{5}{6}$
3. 2
4. $-2\frac{1}{2}$
5. $\frac{11}{15}$
6. $-1\frac{3}{4}$
7. 4
8. $-\frac{5}{6}$
9. $-\frac{1}{40}$
10. $-\frac{5}{72}$
11. 4
12. $-\frac{3}{20}$
13. 48.7
14. 54
15. 158
16. 24.4
17. 145.6
18. 99.2
19. 610
20. 70.9
21. 410
22. 694
23. 52.8
24. 1378
25. 150
26. 360
27. –7.65
28. 7 servings

Practice C

1. $2\frac{1}{12}$
2. $\frac{1}{2}$
3. $\frac{2}{3}$
4. $-2\frac{2}{3}$
5. $-\frac{7}{20}$
6. $-\frac{9}{32}$
7. 8
8. 2
9. 63
10. 65.02
11. 3277
12. 9
13. 197,000
14. –2016
15. 15.02
16. –6.45
17. –250
18. –65.2
19. 540
20. $\frac{5}{8}$ yard
21. 42 months

Review for Mastery

1. $\frac{5}{3}$
2. $\frac{1}{6}$
3. $\frac{3}{7}$
4. $\frac{1}{12}; \frac{3}{96}=\frac{1}{32}$
5. $\frac{4}{3}\times\frac{1}{16}=\frac{4}{48}=\frac{1}{12}$
6. $\frac{21}{20}; \frac{3}{4}$
7. $\left(\frac{8}{9}\right); -\frac{2}{3}$
8. $6\overline{)11.4}$; 1.9
9. $3\overline{)45.6}$; 15.2
10. $2\overline{)712}$; 356
11. $8\overline{)5728}$; 716

Challenge

1. –4.75
2. 4.5
3. 4
4. –0.5

5. –29

6. 0

7. 24

8. 3

9. $\frac{9}{25}$

10. $-\frac{3}{8}$

11. $\frac{1}{36}$

12. $\frac{1}{10,000}$

13. 1

14. 3

15. By definition, ↲100–digit number↲= 99. Then, since 99 has 2 digits, ↲99↲=1.

Problem Solving

1. 0.005 hours
2. 0.006 hours
3. 0.009 hours
4. 1.471 hours
5. 1.667 hours
6. A
7. G
8. D
9. F

Reading Strategies

1. an exchange
2. $\frac{8}{7}$
3. $\frac{5}{6}$
4. $\frac{7}{7}$, or 1
5. 1
6. $\frac{2}{1}$, or 2
7. 1

Puzzles, Twisters & Teasers

25.4; 2; 16.8; 32; 12.4; 3; 17.6; 2.8; 312; 1630

L U N C H A N D

D I N N E R

LESSON 2-6

Practice A

Possible answers:

1. 4
2. 9
3. 24
4. 6
5. $\frac{7}{10}$
6. $1\frac{7}{12}$
7. $\frac{1}{2}$
8. $-\frac{5}{14}$
9. $2\frac{1}{6}$
10. $\frac{3}{4}$
11. $1\frac{5}{9}$
12. $\frac{9}{10}$
13. $-\frac{1}{10}$
14. $3\frac{5}{6}$
15. $1\frac{5}{12}$
16. $-\frac{1}{6}$
17. $-\frac{7}{8}$
18. $-\frac{1}{2}$
19. $\frac{3}{8}$
20. $1\frac{1}{9}$
21. $\frac{1}{2}$
22. $-\frac{5}{14}$
23. $7\frac{7}{8}$ ft

Practice B

1. $1\frac{1}{6}$
2. $\frac{14}{15}$
3. $\frac{5}{12}$
4. $-\frac{1}{18}$
5. $-\frac{5}{16}$
6. $1\frac{11}{18}$
7. $\frac{5}{8}$
8. $\frac{11}{24}$
9. $6\frac{7}{24}$
10. $3\frac{5}{18}$
11. $2\frac{1}{15}$
12. $-\frac{11}{12}$
13. $4\frac{7}{9}$
14. $7\frac{1}{4}$
15. $4\frac{5}{6}$
16. $-1\frac{9}{10}$
17. $4\frac{5}{24}$
18. $-\frac{1}{15}$

19. $\frac{9}{70}$

20. $-\frac{13}{24}$

21. $-\frac{7}{12}$

22. $\frac{1}{5}$

23. $19\frac{5}{12}$ h

Practice C

1. $1\frac{5}{36}$
2. $\frac{29}{30}$
3. $-\frac{1}{24}$
4. $\frac{25}{32}$
5. $-3\frac{41}{72}$
6. $4\frac{11}{12}$
7. $2\frac{31}{50}$
8. $-2\frac{1}{10}$
9. $7\frac{16}{25}$
10. $\frac{5}{9}$
11. $-4\frac{13}{21}$
12. $-\frac{3}{22}$
13. $4\frac{23}{30}$
14. $-\frac{19}{27}$
15. $3\frac{15}{16}$ ft
16. down $2\frac{31}{32}$

Review for Mastery

Possible model.

1.

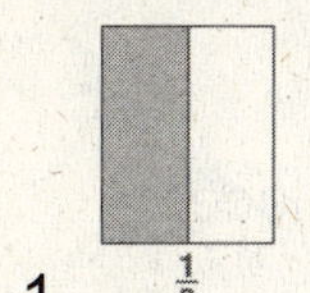

$\frac{1}{2}$

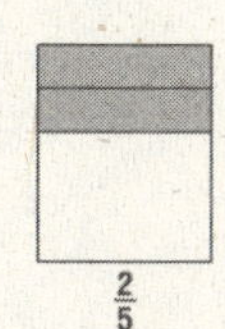

$\frac{2}{5}$

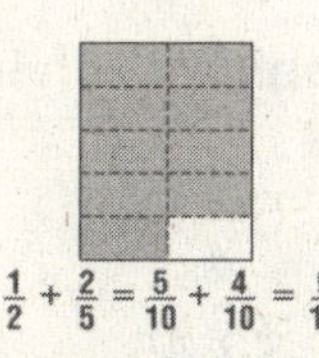

$\frac{1}{2}+\frac{2}{5}=\frac{5}{10}+\frac{4}{10}=\frac{9}{10}$

Possible model.

2.

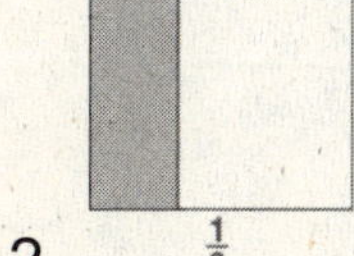

$\frac{1}{3}$

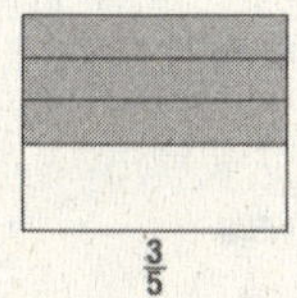

$\frac{3}{5}$

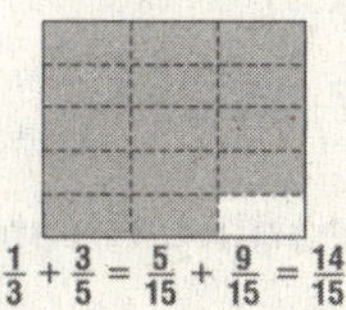

$\frac{1}{3}+\frac{3}{5}=\frac{5}{15}+\frac{9}{15}=\frac{14}{15}$

3. 12; 6, (12); 4, 8, (12)
4. 21; 3, 6, 9, 12, 15, 18, (21); 7, 14, (21)
5. 5; 12; $\frac{17}{20}$

6. 12; 5; $\frac{17}{16}$; $1\frac{1}{16}$
7. 8; 15; $7\frac{23}{24}$
8. $\frac{3}{5}$
9. $\frac{11}{45}$
10. $\frac{17}{60}$

Challenge

1. $0.0\overline{1}$
2. $0.0\overline{21}$
3. $0.0\overline{358}$
4. $0.0\overline{4}$
5. $0.0\overline{62}$
6. $0.0\overline{617}$
7. $\frac{7}{9}$
8. $\frac{8}{90}=\frac{4}{45}$
9. $\frac{24}{9900}=\frac{2}{825}$
10. $\frac{2}{10}+\frac{8}{90}=\frac{26}{90}=\frac{13}{45}$
11. $\frac{25}{100}+\frac{32}{9900}=\frac{2507}{9900}$
12. $\frac{1}{10}+\frac{27}{990}=\frac{126}{990}=\frac{7}{55}$
13. $\frac{75}{100}+\frac{483}{99,900}=\frac{75,408}{99,900}=\frac{6284}{8325}$

Problem Solving

1. $102\frac{3}{8}$ inches
2. $53\frac{7}{8}$ inches
3. $\frac{7}{8}$ cup
4. $7\frac{2}{5}$ inches
5. D
6. H

Reading Strategies

1. 2 and 4
2. 2
3. $\frac{1}{2}=\frac{2}{4}$
4. $\frac{2}{4}+\frac{1}{4}=\frac{3}{4}$
5. 3, 6
6. 2; $\frac{2}{6}$
7. $\frac{5}{6}-\frac{2}{6}=\frac{3}{6}=\frac{1}{2}$

Puzzles, Twisters & Teasers

$\frac{71}{84}$; $\frac{44}{45}$; $\frac{3}{80}$; $1\frac{5}{12}$; $4\frac{4}{45}$; $\frac{19}{24}$; $\frac{67}{112}$;

$-2\frac{4}{9}$; $\frac{7}{16}$; $-2\frac{7}{60}$

IN THE

CROAK ROOM

LESSON 2-7

Practice A

1. $x = 3.4$
2. $a = 8.4$
3. $m = 2$
4. $x = 2.6$
5. $w = -7.8$
6. $n = -7.22$
7. $y = -8$
8. $k = 10.25$
9. $d = 12$
10. $x = \frac{4}{5}$
11. $x = 2$
12. $a = 2\frac{1}{4}$
13. $x = 1\frac{7}{10}$
14. $x = -\frac{2}{7}$
15. $a = -\frac{3}{4}$
16. $12\frac{3}{4}$ in.
17. $119.25

Practice B

1. $x = 5.39$
2. $y = 15.54$
3. $w = 125$
4. $a = 74.21$
5. $x = 12.35$
6. $d = 85.68$
7. $m = -1.7$
8. $n = -3.56$
9. $x = -38.772$
10. $x = \frac{1}{3}$
11. $y = -1\frac{1}{2}$
12. $d = 1\frac{1}{2}$
13. $x = \frac{1}{14}$
14. $x = 3\frac{4}{7}$
15. $a = -1\frac{11}{16}$
16. 7 cups of flour and $3\frac{3}{4}$ cups of sugar
17. 19.5 mi

Practice C

1. $x = -13.74$
2. $m = 41.4$
3. $w = 1976.022$
4. $x = -2\frac{1}{45}$
5. $a = \frac{3}{5}$
6. $y = 5\frac{1}{80}$
7. $10\frac{1}{2}$ in. by $15\frac{3}{4}$ in.
8. 7.46 mi
9. $8.60
10. 16.5 in

Review for Mastery

1. add 1.4
2. subtract $\frac{1}{4}$
3. divide by 3
4. $x = -9.25$
5. $y = 81$
6. $5860.92 = z$
7. $x = -55$
8. $125 = k$
9. $m = 1\frac{4}{15}$

Challenge

1. $x = 21$
2. $x = 2$
3. $x = 1\frac{1}{2}$

Problem Solving

1. $5\frac{1}{4}$ feet
2. 31 pieces
3. $2.45
4. –252.87 ºC
5. C
6. G
7. A
8. G

Reading Strategies

1. Get y by itself on one side of the equation.
2. subtraction
3. $\frac{1}{4} - \frac{1}{4} + y = \frac{3}{4} - \frac{1}{4}$
4. $y = \frac{2}{4}$ or $\frac{1}{2}$

5. Get x by itself on one side of the equation.
6. addition
7. $x - 4.5 + 4.5 = 13 + 4.5$
8. $x = 17.5$

Puzzles, Twisters & Teasers

1. $\frac{2}{3}$
2. $\frac{5}{12}$
3. $\frac{1}{2}$
4. $-\frac{2}{3}$
5. 54.7
6. −2.4
7. 0
8. 6.3
9. −19.2
10. $-\frac{2}{9}$
11. −3
12. −5.4

TOO MANY PROBLEMS

LESSON 2-8

Practice A

1. subtract 2; divide by 3
2. add 1; multiply by 4
3. $x = 3$
4. $x = 18$
5. $a = -1$
6. $x = -8$
7. $y = 6$
8. $x = 7$
9. $w = -3$
10. $r = -1$
11. x = # of miles;
 $45 + 0.25x = 66$;
 $45 - 45 + 0.25x = 66 - 45$;
 $0.25x = 21$;
 $\frac{0.25x}{0.25} = \frac{21}{0.25}$;
 $x = 84$ miles

Practice B

1. x = # of uniforms;
 $1762 = 598 + 24.25x$;
 $1762 - 598$;
 $= 598 - 598 + 24.25x$;
 $1164 = 24.25x$;
 $\frac{1164}{24.25} = \frac{24.25x}{24.25}$;
 $48 = x$
2. x = # of laps
 $4 = \frac{3}{4} + \frac{1}{4}x + \frac{3}{4}$
 $4 - \frac{6}{4} = \frac{6}{4} - \frac{6}{4} + \frac{1}{4}x$
 $\frac{5}{2} = \frac{1}{4}x$
 $4\left(\frac{5}{2}\right) = \left(\frac{1}{4}x\right)4$
 $10 = x$
3. $a = 31$
4. $x = -10$
5. $y = -14$
6. $k = 55$
7. $x = 4$
8. $x = -14$
9. $n = 15$
10. $a = -1$
11. $x = 3$
12. $r = -6.1$
13. $w = -4$
14. $r = -1.22$
15. 31 minutes
16. 11

Practice C

1. $5x + 3 = 28$; $x = 5$
2. $18 - 4x = 62$; $x = -11$
3. $\frac{3x + 7}{5} = 17$; $x = 26$
4. $\frac{x}{5} - 2 = 8$; $x = 50$
5. $x = 9$
6. $w = 4\frac{1}{2}$
7. $x = -25$
8. $n = 39$
9. $x = -1.5$
10. $y = 5$
11. $n = 3$
12. $a = 10$
13. $d = 28\frac{1}{2}$
14. $r = -3.5$
15. $x = -4$
16. $a = 2$
17. 9 weeks
18. $196

Review for Mastery

1. multiplied by 3; 5 is subtracted; add 5; divide by 3
2. divided by 3; 5 is added; subtract 5; multiply by 3
3. added to the variable; divided by 3; multiply by 3; subtract 5
4. multiplied by –3; 2 is subtracted; add 2; divide by –3
5. subtracted from; divided by 5; multiply by 5; add 2
6. –7; –7; 12; 3; 12; 3; 4;

   Check: 4; 12; 19 = 19
7. +7; +7; 12; 3; 3; 36;

   Check: 36; 12; 5 = 5
8. 2; 2; 16; +3; +3; 19;

   Check: 19; 16; 8 = 8

Challenge

1. Let x = number of lime candies.

   $\frac{x}{36} = \frac{4}{9}$;

   $36 \cdot \frac{x}{36} = \frac{4}{9} \cdot 36$;

   $x = 16$; 16
2. Let x = number of lemon cookies. Then $x - 7$ = number of mint cookies.

   $\frac{x}{64} = \frac{5}{8}$;

   $64 \cdot \frac{x}{64} = \frac{5}{8} \cdot 64$;

   $x = 40 \leftarrow$ lemon cookies; 33

Problem Solving

1. 201 min
2. 184 min
3. 200 min
4. Plan A
5. Plan B
6. B
7. J
8. A
9. F

Reading Strategies

1. \$20
2. \$8
3. \$20 – \$8 = \$12
4. \$12
5. \$45
6. \$6
7. \$45 – \$6 = \$39
8. \$39 ÷ 3 = \$13
9. \$13

Puzzles, Twisters & Teasers

1. –55
2. 23
3. 8
4. –4
5. –6.4
6. 7
7. 3
8. –5
9. 4
10. –30

A T T H E S N O W B A L L